RECENT ADVANCES IN ARTIFICIAL NEURAL NETWORKS
Design and Applications

The CRC Press

International Series on Computational Intelligence

Series Editor
L.C. Jain, Ph.D., M.E., B.E. (Hons), Fellow I.E. (Australia)

L.C. Jain, R.P. Johnson, Y. Takefuji, and L.A. Zadeh
Knowledge-Based Intelligent Techniques in Industry

L.C. Jain and C.W. de Silva
Intelligent Adaptive Control: Industrial Applications in the Applied Computational Intelligence Set

L.C. Jain and N.M. Martin
Fusion of Neural Networks, Fuzzy Systems, and Genetic Algorithms: Industrial Applications

H.-N. Teodorescu, A. Kandel, and L.C. Jain
Fuzzy and Neuro-Fuzzy Systems in Medicine

C.L. Karr and L.M. Freeman
Industrial Applications of Genetic Algorithms

L.C. Jain and B. Lazzerini
Knowledge-Based Intelligent Techniques in Character Recognition

L.C. Jain and V. Vemuri
Industrial Applications of Neural Networks

H.-N. Teodorescu, A. Kandel, and L.C. Jain
Soft Computing in Human-Related Sciences

B. Lazzerini, D. Dumitrescu, L.C. Jain, and A. Dumitrescu
Evolutionary Computing and Applications

B. Lazzerini, D. Dumitrescu, and L.C. Jain
Fuzzy Sets and Their Application to Clustering and Training

L.C. Jain, U. Halici, I. Hayashi, S.B. Lee, and S. Tsutsui
Intelligent Biometric Techniques in Fingerprint and Face Recognition

Z. Chen
Computational Intelligence for Decision Support

L.C. Jain
Evolution of Engineering and Information Systems and Their Applications

RECENT ADVANCES IN ARTIFICIAL NEURAL NETWORKS
Design and Applications

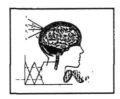

Edited by
Lakhmi Jain, Ph.D.
University of South Australia

Anna Maria Fanelli, Ph.D.
University of Bari, Italy

CRC Press
Taylor & Francis Group
Boca Raton London New York

CRC Press is an imprint of the
Taylor & Francis Group, an **informa** business

First published 2000 by CRC Press
Taylor & Francis Group
6000 Broken Sound Parkway NW, Suite 300
Boca Raton, FL 33487-2742

First issued in paperback 2020

Reissued 2018 by CRC Press

Library of Congress Cataloging-in-Publication Data

Recent advances in artificial neural networks : design and applications / edited by
Lakhmi C. Jain, Anna Maria Fanelli
        p. cm — (The CRC Press international series on computational intelligence)
    Includes bibliographical references and index.
    ISBN 0-8493-2268-5 (alk. paper)
    1. Neural networks (Computer science) I. Jam, L. C. II. Fanelli, Anna Maria
  III. Series.
  QA 76.87 .R43 2000   99058848
  006.3'2—dc21

A Library of Congress record exists under LC control number: 99058848

ISBN 13: 978-0-367-57246-4 (pbk)
ISBN 13: 978-1-315-89711-0 (hbk)

Visit the Taylor & Francis Web site at http://www.taylorandfrancis.com and the
CRC Press Web site at http://www.crcpress.com

PREFACE

Neural networks are a new generation of information processing paradigms designed to mimic some of the behaviors of the human brain. These networks have gained tremendous popularity due to their ability to learn, recall and generalize from training data. A number of neural network paradigms have been reported in the last four decades, and in the last decade the neural networks have been refined and widely used by researchers and application engineers.

The main purpose of this book is to report recent advances in neural network paradigms and their applications. It is impossible to include all recent advances in this book; hence, only a sample has been included.

This book consists of 10 chapters. Chapter 1, by Ghosh and Taha, presents the architecture of a neuro-symbolic hybrid system. This system embeds initial domain knowledge and/or statistical information into a custom neural network, refines this network using training data, and finally extracts refined knowledge in the form of refined rule base. Two successful applications of this hybrid system are described.

Chapter 2, by Karayiannis and Behnke, presents an axiomatic approach for formulating radial basis function neural networks. The batch and sequential learning algorithms are developed for reformulated radial basis function neural networks. This approach is demonstrated on handwritten digit recognition.

Chapter 3, by Vassilas, is on efficient neural network-based methodology for the design of multiple classifiers. An increase in speed in the neural network training phase as well as in the selection of fuzzy and statistical supervised classifiers is achieved by size reduction and redundancy removal from the data set. The catalog of self-organizing feature maps together with the index table is used as a compressed representation of the original data. This technique is demonstrated on land-cover classification of multi-spectral satellite image showing increased speed.

Versino and Gambardella describe the design of a self-organizing map-like neural network which learns to associate actions with perceptions under the supervision of a planning system in Chapter 4. This novel technique is validated in learning fine motion in robotics.

Chapter 5, by Fernández-Delgado, Presedo, Lama, and Barro, is on a new neural network called MART for adaptive pattern recognition of multichannel input signals. A real application related to the multichannel signal processing is presented to demonstrate the ability of this network to solve complex problems.

Caudell and Healy present their research on a new version of the lateral priming adaptive resonance theory network in Chapter 6. They demonstrate that this architecture not only has one of the highest bounds on learning convergence, but also has strong empirical evidence of excellent generalization performance.

Chapter 7, by Aboulenien and De Wilde, discusses an intelligent agent that employs a machine learning technique in order to provide assistance to users dealing with a particular computer application. The authors present actual results from a prototype agent built using this technique applied on flight reservation domain.

Chapter 8, by Halici, Leblebicioglu, Özgen, and Tuncay, presents some applications of neural networks in process control. The authors show that hybrid methods using neural networks are very promising for the control of nonlinear systems.

Chapter 9, by Howlett, de Zoysa, and Walters, is on monitoring internal combustion engines by neural network based virtual sensing. It is necessary to reduce the quantities of polluting gases emitted by this engine and to decrease the amount of fuel consumed per kilometer. The use of neural networks for monitoring the parameters of this engines is proposed.

Pedrycz presents a novel approach to pattern classification using a concept of fuzzy Petri nets in Chapter 10. The learning scheme is illustrated with the aid of numeric examples.

This book will be useful for application engineers, scientists, and research students who wish to use neural networks for solving real-world problems.

We would like to express our sincere thanks to Berend-Jan van der Zwaag, Irene van der Zwaag-Tong, Ashlesha Jain, Ajita Jain and Sandhya Jain for their help in the preparation of the manuscript. We are grateful to the authors for their contributions, and thanks are due to the CRC press for editorial assistance.

Lakhmi Jain, Australia
Anna Maria Fanelli, Italy

The Editors

Lakhmi Jain is a Director/Founder of the Knowledge-Based Intelligent Engineering Systems (KES) Centre, located in the University of South Australia. He is a fellow of the Institution of Engineers Australia. He has initiated a postgraduate stream by research in the Knowledge-based Intelligent Engineering Systems area. He has presented a number of Keynote addresses in International Conferences on Knowledge-Based Systems, Neural Networks, Fuzzy Systems and Hybrid Systems.

He is the Founding Editor-in-Chief of the International Journal of Knowledge-Based Intelligent Engineering Systems and served as an Associate Editor of the IEEE Transactions on Industrial Electronics. Dr Jain was the Technical chair of the ETD2000 International Conference in 1995, and Publications Chair of the Australian and New Zealand Conference on Intelligent Information Systems in 1996. He also initiated the First International Conference on Knowledge-based Intelligent Electronic Systems in 1997. This is now an annual event. He served as the Vice President of the Electronics Association of South Australia in 1997. He is the Editor-in-Chief of the International Book Series on Computational Intelligence, CRC Press USA. His interests focus on the applications of novel techniques such as knowledge-based systems, artificial neural networks, fuzzy systems and genetic algorithms and the application of these techniques

Anna Maria Fanelli received the "Laurea" degree in Physics from the University of Bari, Italy, in 1974. From 1975 to 1979, she was full time researcher at the Physics Department of the University of Bari, Italy, where she became Assistant Professor in 1980. In 1985 she joined the Department of Computer Science at the University of Bari, Italy, as Professor of Computer Science. Currently, she is responsible for the courses "Computer Systems Architectures" and "Neural Networks" in the computer science discipline Her research activity has involved issues related to pattern recognition, image processing and computer vision. Her work in these areas has been published in several journals and conference proceedings. Her current research interests include artificial neural networks, genetic algorithms, fuzzy systems, neuro-fuzzy modeling and hybrid systems.

Professor Fanelli is a member of the IEEE Society, the International Neural Network Society and AI*IA (Italian Association for Artificial Intelligence). She is the editorial board member of the *International Journal of Knowledge-Based Intelligent Engineering Systems.*

CONTENTS

Chapter 2.
New radial basis neural networks and their application in a large-scale handwritten digit recognition problem.........39
N.B. Karayiannis and S. Behnke

Chapter 3.

Efficient neural network-based methodology for the design of multiple classifiers.. 95
N. Vassilas

Chapter 4.

Learning fine motion in robotics: design and experiments. 127
C. Versino and L.M. Gambardella

Chapter 5.
A new neural network for adaptive pattern recognition of multichannel input signals 155
M. Fernández-Delgado, J. Presedo, M. Lama, and S. Barro

Chapter 6.
Lateral priming adaptive resonance theory (LAPART)-2: innovation in ART 187
T.P. Caudell and M.J. Healy

Chapter 9.

Monitoring internal combustion engines by neural network based virtual sensing............ 291

R.J. Howlett, M.M. de Zoysa, and S.D. Walters

Chapter 10.
Neural architectures of fuzzy Petri nets.............................319
W. Pedrycz

Index   347

This book is dedicated to all our students

CHAPTER 1

A NEURO-SYMBOLIC HYBRID INTELLIGENT ARCHITECTURE WITH APPLICATIONS

J. Ghosh
Department of Electrical and Computer Engineering
University of Texas
Austin, TX 78712-1084
U.S.A.
ghosh@ece.utexas.edu

I. Taha
Military Technical College
Cairo, Egypt
ismail.taha@mailexcite.com

Hybrid Intelligent Architectures synergistically combine the strengths of diverse computational intelligence paradigms and avail of both domain knowledge and training data to solve difficult learning tasks. In particular, several researchers have studied some aspects of combining symbolic and neural/connectionist approaches, such as initializing a network based on existing rules, or extracting rules from trained neural networks. In this chapter, we present a *complete system* that embeds initial domain knowledge and/or statistical information into a custom neural network, refines this network using training data, and finally extracts back refined knowledge in the form of a refined rule base with an associated inference engine. Two successful applications of this hybrid architecture are described.

1   Introduction

The synergistic use of multiple models to difficult problems has been advocated in a variety of disciplines. Such approaches can yield systems that not only perform better, but are also more comprehensive and *robust*. A strong motivation for such systems was voiced by Kanal in his classic 1974 paper [17], prompting work on combining linguistic and statistical models, and heuristic search with statistical pattern recognition. In nonlinear control, multiple model methods, such as gain-schedule control, have a long tradition (see http://www.itk.ntnu.no/ansatte/Johansen_Tor.Arne/mmamc/address.html for a detailed list of researchers).

Sentiments on the importance of multiple approaches have also been voiced in the AI community, for example, by Minsky [27]:

> " To solve really hard problems, we'll have to use several different representations.... It is time to stop arguing over which type of pattern-classification technique is best.... Instead we should work at a higher level of organization and discover how to build managerial systems to exploit the different virtues and evade the different limitations of each of these ways of comparing things."

Indeed, there are several examples of successful multi-model approaches in the "learning" community – from the theory of neural network ensembles and modular networks [31] to multistrategy learning [26]. Hybridization in a broader sense is seen in efforts to combine two or more of neural network, Bayesian, GA, fuzzy logic and knowledge-based systems [1], [4], [25], [35]. The goal is again to incorporate diverse sources and forms of information and to exploit the somewhat complementary nature of different methodologies.

The main form of hybridization of interest in this chapter involves the integration of symbolic and connectionist approaches [6], [8], [13], [15], [18], [24], [35], [41]. Such combinations have attracted widespread interest for several reasons that are founded on the complementary strengths and weaknesses of these two approaches. For example, in many application domains, it is hard to acquire the complete domain knowledge and represent it in a rule based format. Moreover, the acquired knowledge may be uncertain or inconsistent [16], [30]. Expert systems can also suffer from the brittleness of rules and lead to problems when the domain theory is noisy [33]. Data driven connectionist models, on the other hand, can be trained in a supervised or unsupervised fashion to perform reasonably well even in domains with noisy training data. However, they cannot as readily incorporate existing domain knowledge or provide a symbolic explanation of results. Finally we note that in many real life situations, there is some amount of existing domain knowledge (which a purely model free neural network cannot exploit) as well as domain data that may be acquired over time. Being able to use both knowledge and data is paramount, specially in "non-stationary" scenarios that demand continuous model tuning or refinement, thus further motivating hybrid methods.

In this chapter, we present a comprehensive **Hybrid Intelligent Architecture (HIA)** that augments a knowledge base system with connectionist and statistical models to help the former refine its domain knowledge and improve its performance and robustness. Figure 1 shows the key modules of HIA.

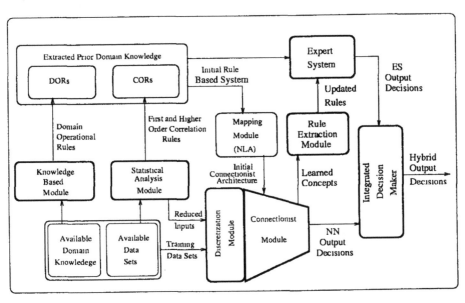

Figure 1. Major components of the Hybrid Intelligent Architecture.

The (optional) rule based system represents the initial domain theory extracted from domain experts in a rule-based format. The acquired rules are mapped into an initial connectionist architecture with uniform structure. The (optional) statistical module analyzes the available data sets and extracts certain correlations between different input parameters and also between input parameters and output decisions. The extracted statistical information is used to provide the mapped initial connectionist architecture with first and higher order input-input and input-output correlation rules. It is also used to provide supplementary rules to an initial rule-based system.

Before training the initial connectionist architecture, a fuzzy subsystem incorporating a coarse coding scheme is used to discretize the input parameters into multi-interval inputs with initial mean and variance for each interval. During the training phase of the connectionist architecture, an Augmented Backpropagation Algorithm (ABA) with momentum term is used to refine the discretization parameters and thus enhance the domain parameters. Therefore, the connectionist architecture can improve the efficiency of the domain theory and incorporate it in its topology.

At the end of the training phase, the final connectionist architecture, with the updated weights and links, can be viewed as a revised domain theory. It can be used to update the initial expert system with new learned concepts. Moreover, it can be converted back, if needed, to a rule based format to achieve the power of explanation [2], [12], [40]. Furthermore, one can use an integrated decision maker to combine the decisions taken by the updated expert system and the trained connectionist architecture and provide the combined decisions to the user.

The rest of this chapter describes in detail the different modules of HIA. In Sections 2-6, we elaborate upon the concepts of discretizing continuous inputs into multi-interval inputs, mapping available domain knowledge into a connectionist architecture, and enhancing the discretization parameters during the training phase. Sections 7-9 summarize options for rule extraction and output integration. Sections 10 and 11 describe two applications of HIA: (1) controlling water reservoirs of the Colorado river around Austin, and (2) characterizing and classifying the Wisconsin Breast Cancer Data Set. In the concluding section we comment on the relation between HIA and some other hybrid and fuzzy approaches, summarize the significance of the current work and point to future directions. Further details on HIA can be found in [34].

2   Knowledge Based Module for Representation of Initial Domain Knowledge

As depicted in the bottom left of Figure 1, available domain information can be divided into two parts: the knowledge that represents the operational rules of the domain and the data sets that represent the historical records of the application domain.

The first module in HIA is a knowledge based module that is used to represent the initial domain knowledge through a well-defined format named *Domain Operational Rule (DOR)* format. The DORs are built using only the basic domain primitives that can be acquired easily from the domain without consuming much time or effort. The basic components needed to build the DORs are: (i) the domain objects and their attributes; (ii) the relationship between the domain objects; and (iii) the valid range of the attributes. These basic components represent the initial domain theory and may not be sufficient for representing the complete problem in a rule-based format. However, they can be used to build an initial rule-based expert system. The **DOR** format is a general rule-based format that can be used to represent rule-based systems with and without certainty factors. In case of rule-based systems without certainty factors, the value of cf is replaced by "1" in each rule. The following rules describe the syntax of the DOR format using the Backus-Naur Form (BNF):

If *Compound-Condition* [*OR Compound-Condition*]* $\xrightarrow{cf}$ *Consequent*$^+$

Compound-Condition ::= *Simple-Condition* | *Simple-Condition* "AND" *Compound-Condition*

Simple-Condition ::= [*NOT*] *Boolean-Expression*

Consequent ::= *Output-Variable*

where the symbol ::= means *"to be written as"*, | a vertical bar to represent choices, [·] is an optional term that can be repeated one time, [·]* is an optional term that can be repeated zero or more times, and [·]$^+$ is a term that can be repeated one or more times.

In many real applications, rules are not always fully true. Therefore, each rule represented in the DOR format has an attached certainty factor value, cf, which indicates the measure of belief, or disbelief if it is negative, in the rule consequent provided the premises (left hand side) of the rule are true. It is important to mention that:

- Rule consequents in the DOR format are not permitted to be used as conditions in any other rule. Such a restriction was introduced to avoid increasing the number of hidden layers of the connectionist architecture and hence reduces its complexity. This restriction leads to a simpler uniform connectionist network, as seen later.

- In spite of this restriction, the DOR format can be used for rule-based systems that allow rule consequents to be used as conditions (premises) in other rules. Such systems can be represented in the DOR format by replacing the condition, say of rule R_n, that has appeared as a consequent in another rule, say rule R_m, by the left hand side of the latter rule (R_m).

- The DOR format does not have any restriction on the number of conditions per rule or the order of logical operators in any of its rules.

- The rules represented in the DOR format are mapped directly into a corresponding initial connectionist architecture without any limitation on the number or the order of the operators in each rule. Another approach of mapping domain operational rules into an initial connectionist architecture is to convert them to another set of rules with only conjuncted (anded) premises to simplify the mapping operation [8]. The latter approach simplifies the mapping phase but it increases the complexity of the mapped connectionist architecture.

3   Extraction of Supplementary Rules via the Statistical Analysis Module

In many application domains, extracting complete domain operational rules from domain representatives suffers from the knowledge acquisition bottleneck [16], [30]. Therefore, we need to seek another source of information to get more prior knowledge from the application domain, such as statistical information from available datasets. We have investigated two simple statistical approaches to extract prior domain knowledge from the available data sets. As shown in Figure 2, the first approach is to extract the first and the higher order correlation coefficients of the available data sets to generate supplementary and constraint rules to the DORs extracted by the knowledge based module. The second approach is to project the input features vector into a lower dimensional input space and only deal with the most intrinsic input features [7], [14]. These two approaches can be done independently and their results can be combined with the extracted initial DORs. The following subsection presents how supplementary correlation rules can be extracted from available data sets and then used to update the initial rule-based system. Then the next subsection presents how input features can be projected into a lower dimensional input space.

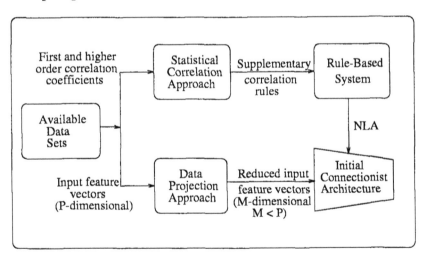

Figure 2. Two statistical approaches to extract additional domain knowledge.

3.1   Extraction of Correlation Rules

The correlation approach of the statistical module analyzes available data sets by extracting *"certain"* correlation rules between each pair of inputs and also between each input-output parameter. In addition, it extracts the main statistics of each input and output parameter (e.g., the mean and the variance). These statistics are used later to initialize the adaptive parameters of the discretization module.

Assume that X represents an input feature vector and Y represents an output decision vector; let $CC_X = \frac{CO(X)}{\sqrt{\sigma_X}}$ and $CC_{XY} = \frac{CO(X,Y)}{\sqrt{\sigma_X \sigma_Y}}$, where $CO(\cdot)$ represents the covariance matrix. Thus CC_X is the input correlation coefficients matrix and CC_{XY} represents the cross-correlation coefficients matrix between the input and the output vectors X and Y. After computing CC_X and CC_{XY}, a threshold value, θ_0, is chosen based on the application domain (usually $\theta_0 \geq 0.80$). Based on the chosen threshold θ_0 and the elements of the correlation coefficients matrices CC_X and CC_{XY}, the statistical module starts generating COrrelation Rules, named CORs. The statistical analysis module extracts input-input and input-output CORs based on the following algorithm:

1. Let c_{ij} be the correlation coefficient between any two pair of parameters X_i and X_j.

2. **IF** $c_{ij} \geq \theta_0$
   THEN create the COR R_1:

$$IF \quad X_i \quad \xrightarrow{w_1} \quad X_j \quad\quad\quad (R_1)$$

   where the value of w_1 represents the confidence level of the generated rule and $w_1 = cij$.

3. **IF** $c_{ij} \leq -\theta_0$
   THEN create the COR R_2:

$$IF \quad NOT \quad X_i \quad \xrightarrow{w_2} \quad X_j \quad\quad\quad (R_2)$$

   where the value of w_2 equals the magnitude of c_{ij}.

4. **IF** $-\theta_0 > c_{ij} < \theta_0$
   THEN no rule is generated and the statistical module at this point can not conclude any certain correlation between these two parameters.

The CORs generated by the statistical module are represented in the same DOR format to match the initial rule-based module extracted by the knowledge-based module. Therefore, the CORs and the DORs can be combined together with no additional overhead.

The CORs generated by the statistical module can be used as a constraint or as supplementary rules to the DORs and hence help initializing the connectionist architecture with more prior domain knowledge. The following three cases describe how CORs extracted by the statistical module can be used to simplify, maintain, and support the DORs.

1. **Case1:**

    Assume that the knowledge-based module extracts rule R_3, from the domain experts, and the statistical analysis module extracts rule R_4 based on the correlation between X_1 and X_2.

    $$IF \quad X_1 \quad AND \quad X_2 \xrightarrow{w_3} Y_1 \qquad (R_3)$$
    $$IF \qquad\qquad\qquad X_1 \xrightarrow{w_4} X_2 \qquad (R_4)$$

    Therefore, rule R_4 can be used to simplify the previous DOR R_3 and a new rule R_5 is generated to replace both R_3 and R_4. Note that w_4 should be ≥ 0.80.

    $$IF \qquad\qquad\qquad X_1 \xrightarrow{w_5} Y_1, \qquad (R_5)$$

    where $w_5 = w_3 \times w_4$. Note that the new rule R_5 does not depend on X_2 which is highly correlated with X_1 based on the COR R_4. The logical interpretation of rule R_5 results from combining the semantics of both R_3 and R_4 as follows:

    "IF X_1 is true THEN X_2 is true (with w_4 confidence measure) AND IF X_1 AND X_2 are both true THEN Y_1 is true with a confidence level $= w_3$". This interpretation can be simplified to: *"IF X_1 is true; which implicitly implies that X_2 is true; THEN Y_1 is true with a confidence level $= w_5$"* which represents the semantics of rule R_5.

2. **Case2:**

    If the knowledge-based module extracts rule R_3 and the statistical module extracts rule R_6.

$$IF \qquad NOT \quad X_1 \quad \xrightarrow{.96} \quad X_2 \qquad\qquad (R_6)$$

In this case rule R_3 cannot be fired any more whatever the value of X_1 (i.e., in either cases if X_1 is true or false) because R_6 is considered as a constraint (in this example a strong contradiction) rule to the DOR R_3.

3. **Case3:**
   The statistical module can extract CORs which do not exist in the DORs. As an example, if the statistical module extracts rule R_7

$$IF \qquad\qquad\qquad X_3 \quad \xrightarrow{.89} \quad Y_2 \qquad\qquad (R_7)$$

and there were no other rules in the DORs to represent the logical relationship between X_3 and Y_2. In this case the generated COR R_7 is added to the extracted DORs.

Based on the previous cases, the statistical module can provide the DORs with either a constraint or supplementary CORs. Moreover, if there were no DORs extracted from the application domain, the statistical module is used to generate correlation rules and represent it in the same DOR format. See the experimental results presented in Section 11.

After combining the rules extracted by the knowledge-based and the statistical modules and representing them in the DOR format, the Node-Links Algorithm (NLA) is used to map these combined rules into an initial connectionist architecture.

3.2   Reducing the Input Dimensionality

In many application domains the input data are noisy and may have some redundancy. To obtain better network performance it is important to retain the most intrinsic information in the input data and reduce network complexity at the same time, if it is possible. We use Principal Component Analysis (PCA), a well known technique in multivariate analysis, for this purpose [7], [14].

As a preprocessing step, the correlation matrices represented by CC_X of Equation 1 are used first to determine highly correlated input pairs.

Then, PCA is applied after one variable is removed from each highly correlated pair. This typically enhances the PCA algorithm performance while reducing the connectionist architecture input dimensionality. The resulting feature vector from the PCA algorithm is used as an input to the constructed neural network.

The experimental results presented in Section 11 illustrate how we used the statistics of a public domain data set to extract additional prior domain knowledge.

4   The Mapping Module

The Node-Links Algorithm (NLA) utilizes a set of mapping principles to map the initial domain theory, represented in the DOR/COR format, into an initial connectionist architecture that can learn more new domain concepts during the training phase. It results in a three layer AND-OR tree, as exemplified by Figure 3. Note that a *Negated Simple-Condition* is translated into a negative initial weight (-0.6 or -0.7) for the corresponding link. Also, the NLA creates a hidden node even when there is only one *Simple-Condition* in the premise. This type of hidden node is named *self-anded* hidden node, because it ANDs one input node with itself. Therefore, output nodes are viewed as OR nodes and hidden nodes are viewed either as AND or as *self-anded* nodes. The NLA creates a light link between each *self-anded* node, as well as each AND node, and all other input and output nodes that are not linked with it. Introducing such *self-anded* hidden nodes and light links provides the initial connectionist architecture with the power to learn more domain knowledge and extract new features during the training phase. The overhead due to the introduction of the self-anded nodes and their related links is much less than that incurred by interrupting the training phase and adding, heuristically, a random number of hidden nodes [9]. The initial connectionist architecture generated by the NLA has only three layers, independent of the hierarchy of the initial rule-based system and regardless of the nature of the application domain. Moreover, all hidden nodes functionally implement soft conjunctions and use the sigmoid activation function, which clearly improves the training phase of the initial connectionist architecture. This is in contrast to models that have variable network structure, based on the application domain, and hidden units with different functionalities [6], [10], [44].

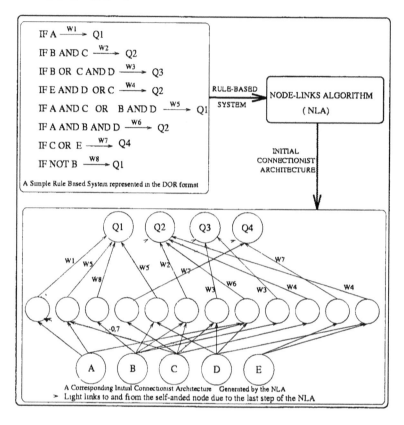

Figure 3. From rule-based system to initial connectionist architecture using the Node-Links Algorithm.

5   The Discretization Module

Measured inputs in control domains are often continuous. Since the operational rules that represent the system use multi-interval ranges to describe the application domain, a discretization function is needed to map continuous inputs into multiple interval inputs. Assume that a continuous measured input z always lies in the range $[a, b]$. A discretization function is used to map it into a corresponding vector $X: (x_1, x_2, ..., x_n)$, where $x_i \in [0, 1]$, $\forall i$ and n is the number of discretized intervals. In a basic symbolic system, exactly one of the x_is is set to 1 for a given z value, and all others are zero. However, we prefer a continuous discretization approach to a binary discretization since it allows "coarse coding," i.e., more than one interval can be active at the same time with different certainty values, based on the value of the measured input z. Coarse coding is a more robust representation of noisy data, which is a prime objective here. This discretization process is a typical fuzzification approach for determining the degree of membership of the measured input z in each interval i. The value of each element x_i is interpreted as the measure of belief that z falls in the ith interval [43].

A Gaussian function with mean μ_i and standard deviation σ_i is selected to represent the distribution of the measured input z over each interval i; so n Gaussian curves are used to fuzzify z into n intervals. [1] The technique is illustrated in Figure 4, where a continuous measured input z gets fuzzified into an input vector X, resulting in $x_1 = 0.25$ and $dx_2 = 0.75$. This fuzzification is done as a preprocessing phase to the initial connectionist architecture. The output of the fuzzification process, X, represents the activation values of the input nodes of the initial connectionist architecture, where each interval is represented by an input node. If the application domain has k continuous measured inputs the fuzzification approach results in a total of $\sum_{i=0}^{k} n_k$ input nodes, where n_k is the number of discretized intervals of the kth measured input.

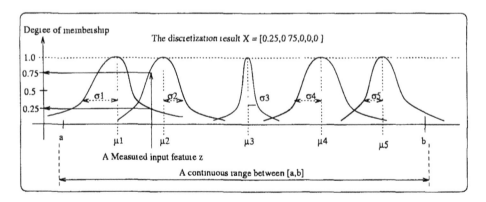

Figure 4.   Discretizing a continuous measured input into n intervals using n Gaussian functions.

---

[1]The choice of the differentiable Gaussian function instead of the typical triangular membership functions used in fuzzy logic is important as it facilitates membership adaptation, as described in the next section.

6   Refining Input Characterization

The initial connectionist architecture is trained by the output vectors (Xs) of the fuzzification function. Assuming that the measured input values are normally distributed within each interval i with mean μ_i and standard deviation σ_i, the Gaussian functions:

$$x_i = f_i(z) = e^{-\frac{1}{2}(\frac{z-\mu_i}{\sigma_i})^2} \tag{1}$$

are used to discretize the measured input value z. An Augmented version of the Backpropagation Algorithm, **ABA**, with momentum term is used to train the initial architecture and stochastically search for the optimal weights to and from all hidden nodes (anded and self-anded nodes). Moreover, the ABA is used to refine the initial discretization parameters μ_i and σ_i for each interval i. The ABA calculates the stochastic gradient descents of the output error with respect to μ_i and σ_i and propagates them one more step back to the fuzzification function, i.e., to the external inputs of the connectionist architecture. Refining the discretization parameters (μ_i, σ_i) helps the connectionist architecture to extract features from the measured inputs that are more discriminating and thus enhances the decision process. The chain rule was used to derive the derivative of the output error, E, with respect to μ_i and σ_i:

$$\frac{\partial E}{\partial \mu_i} = \frac{\partial E}{\partial f_i(z)} \cdot \frac{\partial f_i(z)}{\partial \mu_i} = \frac{1}{\sigma_i^2} \cdot (z - \mu_i) \cdot \sum_{j=0}^{h-1} w_{ij} \cdot \frac{\partial E}{\partial w_{ij}} \tag{2}$$

$$\frac{\partial E}{\partial \sigma_i} = \frac{\partial E}{\partial f_i(z)} \cdot \frac{\partial f_i(z)}{\partial \sigma_i} = \frac{(z - \mu_i)^2}{\sigma_i^3} \cdot \sum_{j=0}^{h-1} w_{ij} \cdot \frac{\partial E}{\partial w_{ij}} \tag{3}$$

where the term $\sum_{j=0}^{h-1} w_{ij} \cdot \frac{\partial E}{\partial w_{ij}}$ represents the gradient descent of the output error propagated back to all the h hidden nodes linked to i^{th} input node. Note that the $\frac{\partial E}{\partial w_{ij}}$s do not need to be recomputed as they are already obtained from updating the weights into the hidden units. The center and width of the ith interval are adjusted as follows:

$$\mu_{inew} = \mu_{iold} - \eta \cdot \frac{\partial E}{\partial \mu_i} + MomentumTerm \tag{4}$$

$$\sigma_{inew} = \sigma_{iold} - \eta \cdot \frac{\partial E}{\partial \sigma_i} + MomentumTerm \tag{5}$$

7   Rule Extraction

Extraction of symbolic rules from trained neural networks is an important feature of comprehensive hybrid systems, as it helps to:

1. Alleviate the knowledge acquisition problem and refine initial domain knowledge.

2. Provide reasoning and explanation capabilities.

3. Support cross-referencing and verification capabilities.

4. Alleviate the "catastrophic interference" problem of certain ANNs [32]. For models such as MLPs it has been observed that if a network originally trained on one task (data set) is subsequently trained on a different task (statistically different data set), then its performance on the first task degrades rapidly. In situations with multiple operating regimes, one can extract rules before the task or environment changes and thus obtain different rule sets for different environmental conditions. Together with a mechanism for detecting the current environment, this presents one solution to the "context discovery" and "context drift" problems.

Other uses of rule extraction include improving acceptability of the product, transfer of knowledge to a more suitable form, and induction of scientific theories.

The rule extraction module of HIA maps the trained connectionist architecture back into a rule based format. This mapping is much harder than the mapping from an initial rule based system to an initial connectionist architecture because: (i) one should guarantee that the extracted domain concepts should not contradict with certain concepts that are known to be true about the domain, (ii) one should refine uncertain domain concepts, and (iii) new concepts may get extracted. An efficient rule extraction module should be able to deal with these three issues.

Several issues should be carefully considered while designing a rule extraction technique:

1. **Granularity of the explanation feature:** is the level of detailed hypotheses and evidence that the system can provide with each of its output decisions.

2. **Comprehensiveness of the extracted rules:** in terms of the amount of embedded knowledge captured by them. This directly determines the *fidelity* of the extracted rules in faithfully representing the embedded knowledge.

3. **Comprehensibility:** indicated by the number of rules and number of premises in each extracted rule from a trained network.

4. **Transparency of the extracted rules:** in terms of how well the decisions or conclusions can be explained.

5. **Generalization Capability:** on test samples.

6. **Portability:** is the capability of the rule extraction algorithm to extract rules from different network architectures.

7. **Modifiability:** is the ability of extracted rules to be updated when the corresponding trained network architecture is updated or re-trained with different data sets.

8. **Theory Refinement Capability:** that can alleviate the knowledge acquisition bottleneck due to the incompleteness, inconsistency, and/or inaccuracy of initially acquired domain knowledge.

9. **Stability or Robustness:** is a measure of how insensitive the method is to corruptions in the training data or initial domain knowledge.

10. **Complexity and Scalability:** Computational issues that are relevant for large datasets and rule bases.

These issues, in addition to others, should be used to measure the quality and performance of rules extracted from trained neural networks. Note that these issues also depend on the rule representation, insertion and network training methods used. Also, it is difficult to simultaneously optimize all of the above criteria. For example, a very comprehensive technique may extract too many rules, with some of them having many premises, thus degrading the robustness and comprehensibility of the resulting rule base.

A variety of rule-extraction techniques have been proposed in the recent literature [2], [36], [39]. Also see the rule extraction home page at: `http://www.fit.qut.edu.au/~robert/rulex.html`. The methodology behind most of the techniques for rule extraction from MLPs can be summarized in two main steps:

1. For each hidden or output node in the network, search for different combinations of input links whose weighted sum exceeds the bias of the current node.

2. For each of these combination generate a rule whose premises are the input nodes to this combination of links. All premises of a rule are conjuncted.

Either [28], KT [9] and Subset [40] are three notable rule extraction algorithms in this category, which we describe as Link Rule Extraction Techniques.

In this section we summarize three recent techniques for extracting rules from trained feedforward ANNs. The first approach is a binary Black-box Rule Extraction technique. The second and the third approaches belong to the Link Rule Extraction category. Details can be found in [36].

7.1   First Technique (BIO-RE)

The first approach is named **Binarized Input-Output Rule Extraction (BIO-RE)** because it extracts binary rules from any neural network trained with "binary" inputs, based on its input-output mapping. It is surprisingly effective within its domain of applicability. The idea underlying BIO-RE is to construct a truth table that represents all valid input-output mappings of the trained network. BIO-RE then applies a logic minimization tool, Espresso [29], to this truth table to generate a set of optimal binary rules that represent the behavior of the trained networks. For example, an extracted rule: "**IF** Y_1 **AND NOT** Y_2 $\longrightarrow O_1$", is rewritten as "**IF** $X_1 > \mu_1$ **AND** $X_2 \leq \mu_2 \longrightarrow O_1$", where μ_i is set to be the threshold of X_i (see Table 2 for examples). The BIO-RE approach is suitable when the input/output variables are naturally binary or when binarization does not significantly degrade the performance. Also the input size (n) should be small.

7.2 Second Technique (Partial-RE)

The idea underlying Partial-RE algorithm is that it first sorts both positive and negative incoming links for each hidden and output node in descending order into two different sets based on their weight values. Starting from the highest positive weight (say i), it searches for individual incoming links that can cause a node j (hidden/output) to be active regardless of other input links to this node. If such a link exists, it generates a rule: "**IF** $Node_i \xrightarrow{cf} Node_j$", where cf represents the measure of belief in the extracted rule and is equal to the activation value of $node_j$ with this current combination of inputs. If a node i was found strong enough to activate a node j, then this node is marked and cannot be used in any further combinations when checking the same node j. Partial-RE continues checking subsequent weights in the positive set until it finds one that cannot activate the current node j by itself. Partial-RE performs the same procedure for negative links and small combinations of both positive and negative links if the required number of premises in a rule is > 1. Partial-RE algorithm is suitable for large size problems, since extracting all possible rules is NP-hard and extracting only the most effective rules is a practical alternative. See Table 3 for examples.

7.3 Third Technique (Full-RE)

Full-RE first generates intermediate rules in the format:

$$\textbf{IF } [(c_1 \cdot X_1 + c_2 \cdot X_2 + \cdots + c_n \cdot X_n) >= \lambda_j] \xrightarrow{cf} Consequent_j,$$

where: c_i is a constant representing the effect of the ith input (X_i) on $Consequent_j$ and λ_j is a constant determined based on the activation function of node j to make it active. If node j is in the layer above node i then c_i represents the weight value w_{ji} of the link between these two nodes. In cases where the neural network inputs (X_is) are continuous valued inputs, then a range of X_i values may satisfy an intermediate rule, and one would want to determine a suitable extremum value in such a range. To make this tractable, each input range has to be discretized into a small number of values that can be subsequently examined. Thus, each input feature $X_i \in (a_i, b_i)$ is discretized into k intervals [21]. When Full-RE finds more than one discretization value of an input X_i that can satisfy the intermediate rule (i.e., the rule has more than one feasible solution) then it chooses the minimum or the maximum of these values based on the *sign* of the corresponding effect parameter c_i. If c_i is negative then Full-RE chooses the minimum discretization value of X_i; otherwise it chooses the maximum value. However, all selected discretization values should satisfy the left hand side (the inequality) of the intermediate rule and the boundary constraints of all input features of this inequality. Final rules extracted by Full-RE are represented in the same format of Partial-RE except that each μ_i is replaced by one of the discretization boundaries (say $d_{i,l}$) selected by Full-RE as described earlier. See Table 4 for examples.

8   Rule Evaluation and Ordering Procedure for the Refined Expert System

To evaluate the performance of rules extracted from trained networks by any of the three presented techniques (or by any other rule extraction approach), a simple rule evaluation procedure which attaches three performance measures to each extracted rule is developed. The three performance measures used to determine the order of the extracted rules are:

(i) **The soundness measure:** it measures how many times each rule is correctly fired.

(ii) **The completeness measure:** a completeness measure attached to a rule represents how many unique patterns are correctly identified/classified by this rule and not by any other extracted rule that is inspected by the inference engine before this rule. For each extracted set of rules with the same consequent, if the sum of the completeness measures of all rules in this set equals the total number of input patterns having the corresponding output then this set of extracted rules is 100% complete with respect to that consequent. An extracted rule with zero completeness measure but having a soundness measure > 0 means that there is a preceding rule(s), in the order of rule application, that covers the same input patterns that this rule covers. Such a rule may be removed.

(iii) **The false-alarm measure:** it measures how many times a rule is misfired over the available data set. While the values of both the completeness and false-alarm measures depend on the order of rule application and the inference engine the soundness measure does not.

8.1   The Rule Ordering Procedure

An expert system requires a set of rules as well as an inference engine to examine the data, determine which rules are applicable, and prioritize or resolve conflicts among multiple applicable rules. A simple way of conflict resolution is to order the rules, and execute the first applicable rule in this ordering. Finding the optimal ordering of extracted rules is a combinatorial problem. So the following *"greedy"* algorithm to order any set of extracted rules, based on the three performance measures, is developed. The rule ordering algorithm first creates a list L that contains all extracted rules. Assume that the list L is divided into two lists, a head list (L_h) and a tail list (L_t), where L_h is the list of all ordered rules and L_t is the list of all remaining (unordered) rules[2]. Initially, L_h is empty and L_t includes all the extracted rules. A performance criteria is used to select one rule from L_t to be moved to the end of L_h, and the process continues till L_t is null.

The steps of the rule ordering algorithm are as follows:

---

1. Initialize L_h = { }, L_t = {all extracted rules}.

2. **WHILE $L_t \neq$ { }, DO**
   (a) Fire all rules in L_h in order.
   (b) Compute the completeness and false-alarm measures for each rule in L_t using the available data set.
   (c) **IF** *∃ a rule with zero false-alarm*
       THEN *this rule is moved from L_t to the end of L_h[3].*

---

[2]i.e., the ordering of rules in L_t has no effect.
[3]If ∃ more than one rule with zero false-alarm **THEN** select the one with the highest completeness measure out of these rules to be moved from L_t to the end of L_h.

ELSE *Among all rules in L_t select the one with the highest (Completeness - False-alarm) measure; add this rule to the end of L_h, delete it from L_t.*

(d) **IF** $\exists$ *any rule in L_t with a zero completeness measure then remove this rule from L_t. This means that the rules in L_h cover this rule.*

3. **END DO.**

In this chapter, all rules extracted by our approaches are ordered using the above rule ordering algorithm. Also, the measures attached to all extracted rules assume that an inference engine that fires only one rule per input (namely, the first fireable rule) is used.

9   The Integrated Decision Maker

The main objective of combining or integrating different learning modules is to increase the overall generalization capability. Since the set of extracted rules is an *"approximated symbolic representation"* of the embedded knowledge in the internal structure of the corresponding trained network, it is expected that when an input is applied to the extracted rules and the trained network, they will usually both provide the same output decision (see Table 6 for examples). The integration module should be able to choose the *"better"* output decision when the two decisions differ, and to compute the certainty factor of the final output decision. When the two output decisions are different, the integration module can use the following selection criteria to select a suitable decision.

1. Select the sub-system (i.e., the set of extracted rules or the trained ANN) with the highest overall performance if none of the following conditions are satisfied:

2. For any mismatched pair of output decisions, check the value of the neural network output decision (i.e., the activation value of the corresponding output node of the neural net before thresholding).

   (a) If the extracted rule-base is indicated by Rule 1, but the neural network output is significantly high, then choose the neural network instead to provide the final decision. Also, report that the extracted rule-base was not able to identify this case, so that a new rule can be asserted in the current knowledge base to handle such cases in the future.

(b) If the neural network is indicated by Rule 1, but the network output is significantly low, then choose the extracted rule-base instead to provide the final output of this case. Also, report that the neural network was not able to identify this case, so that it can be retrained. This case can also be applied if the difference between the two highest activation values of the neural network output nodes is not significant.

This simple heuristic criterion of selecting one of the two mismatched output decisions was applied for all the three architectures and their corresponding set of extracted rules using the breast cancer problem. The implementation results are given in Table 6.

10   Application: Controlling Water Reservoirs

There are several dams and lakes on the Colorado river near Austin. The decision of specifying the amount of water that should be released from any of these dams and lakes is a complicated process. The Lower Colorado River Authority (LCRA) determines this decision for each dam or lake based on the current elevation of the water in the lake, the inflow rate from upstream dams and lakes, the outflow rate from the current lake, the predicted weather (rain fall rate), the predicted elevation of the downstream dams and lakes, and many other factors.

The LCRA uses upstream and downstream gages to monitor and forecast lake levels. There are two main modes of operation for the Highland Lakes: the first is a daily operation in which downstream demands for water are met by daily releases from Lake Buchanan and Travis to supplement the flow of the lower river. The second is for flood control, which primarily concerns Lake Travis since it is the only reservoir with a dedicated flood pool. When the Colorado river downstream from Highland Lake approaches the warning stage at any of the following downstream gages, the rules of the flood control operating mode are used to determine water release.

We acquired the 18 main operational rules for controlling the flood gates on the Colorado river in the greater Austin area from different documents issued by the LCRA after the 1991 Christmas flood around Austin [19]. Two of the rules regulating the control of Mansfield Dam on Lake Travis, expressed in DOR format, are:

If Projected-Level(Lake-Travis,t0) $>=$ 710 AND
        Projected-Level(Lake-Travis,t0) $<=$ 714 AND
        Projected-Level(Lake-Austin,t1) $<=$ 24.8 AND
        Projected-Level(Bastrop,t2) $<=$ 26.7
    Then
        Open-Up-To 10 Flood-Gates.

If Projected-Level(Lake-Travis,t0) $<=$ 710 AND
        Projected-Level(Lake-Travis,t0) $>=$ 691 AND
        Projected-Level(Lake-Austin,t1) $<=$ 20.5 AND
        Projected-Level(Bastrop,t2) $<=$ 25.5
    Then
        Open-Up-To 6 Flood-Gates.

Lake Austin and Bastrop are downstream from Mansfield Dam, $t0$ is the time when the water level at Lake Travis is measured, and $t1 - t0$ and $t2 - t0$ are the approximate times taken for the released water from Lake Travis to reach Lake Austin and Bastrop respectively (approx. 2 hrs and 24 hrs). All values are measured in feet above mean sea level.

10.1 Implementation Results

The HIA knowledge based module was used to implement the extracted domain knowledge in the DOR format. A data set representing 600 patterns was gathered from the LCRA historical records of the different dams and lakes. The acquired data set was divided into two sets. The first one had 400 patterns and was used as the training set; the second served as the validation set. Each pattern includes the measured elevation of three lakes at some given time and the corresponding best decision based on the extracted domain knowledge represented by the DOR.

The coarse coding scheme described in Section 5 was used to discretize each input (measured elevation z) into a multi-interval input vector (X). Each discretized interval i corresponds to one specific DOR *Simple-Condition* and has an initial mean μ_i and standard deviation σ_i. The Node-Links Algorithm (NLA) was used to map the DORs into a one hidden layer initial connectionist architecture. The resulting architecture has 23 input nodes (representing the different discretized intervals), 18 hidden nodes (representing either AND or *self-anded* soft conjunction concepts), and 8 output nodes representing the possible decisions at any time (i.e., how many flood gates should be opened).

During the training phase, the Augmented Backpropagation Algorithm (ABA) with momentum term stochastically searched the weight space to find the optimal weights and also used the partial derivatives of the output error, as described by Equations 2 through 5, to refine the initial discretization parameters (μ_i and σ_i) of the 23 input nodes. Note that each input node in the mapped architecture corresponds to a discretized interval.

By the end of the learning process, we found that the means (μ) of seven different intervals were shifted and the standard deviations (σ) of four of them were also significantly changed from their initial values. The change in the values of the discretization parameters readily reflects how the ABA exploits the training data set to refine the initial domain knowledge represented by the rule based module. Actually, any change in the discretization parameters directly affects the input of the connectionist architecture and hence its output decisions. Therefore, any refinement in these parameters enhances the output performance of the trained connectionist architecture.

After the training phase, the validation set (200 patterns) was used to observe how the HIA will perform in real flood situations. The decisions taken by the HIA, which uses the ABA for training the mapped connectionist architecture and also for refining domain parameters, were compared with the decisions taken by two different connectionist architectures. The first one is an MLP (with one hidden layer) initialized randomly without any prior domain knowledge and trained by the conventional backpropagation with a momentum term. The second is an MLP (also with one hidden layer) initialized by the same initial rules that were used to initialize the HIA and trained by the conventional backpropagation with a momentum term (i.e., no refinement of discretization parameters was done).

Table 1. Test results of Colorado river problem.

No. of epochs	HIA with refined DOR		MLP initialized randomly		MLP with initial DOR	
	MSE	% match	MSE	% match	MSE	% match
1	0.096	76.176	1.367	64.0	0.344	70.9
10	0.015	93.323	0.889	68.5	0.248	72.6
20	0.014	94.234	0.334	71.7	0.103	74.7
30	0.013	94.234	0.233	72.38	0.094	76.3

The output decisions taken by the three architectures (to specify how many flood gates should be opened at a given time) were compared with the desired decisions that should be taken in each situation based on LCRA operational rules. Table 1 provides a summary of the testing results. The presented results are the average of 10 runs. They represent the performance of each architecture when the test data sets were applied. Two parameters are presented for each architecture, the mean square error and the percentage of patterns for which the maximum output value matched with the desired decision based on the LCRA operational rules. Note that the LCRA selects only one decision per pattern, while the neural network indicates support for each decision by the value of the corresponding output. To compute MSE, we used a target value of "1" for output node corresponding to the desired decision and "0" for all other outputs. The implementation results depicted in Table 1 show that:

- The randomly initialized MLP did not perform well compared with the other two architectures.

- The MLP initialized by the initial domain rules and trained with the conventional backpropagation performed somewhat better than the randomly initialized MLP. However, the performance of this architecture did not improve much on further training.

- The HIA did perform much better than the other two architectures and its performance improved significantly by increasing the number of training epochs.

The improved performance of the HIA is mainly due to the continuous refinement of the discretization parameters during each training epoch.

10.2   Rule Extraction

After the learning phase, the connectionist module of HIA was examined by Partial-RE to find out if any new rules were created during the training. Some of the extracted rules were already there in the initial DORs and others were subsets of existing rules. However, some new and useful rules were also found. For example, in the initial rule based system extracted from the LCRA documents, there was a rule specifying that: if the predicted level of Lake Travis is between 681 and 691 feet msl (mean sea level) then release up to 5,000 cfs (cubic feet per second) if the river, with the release, is no higher than 20.5 ft. at Austin and 25.1 ft. at Bastrop. HIA extracted three new and useful fine rules instead of the previous coarse rule. The first new rule extracted by HIA is: if the predicted level of Lake Travis is between 681 and 683 feet msl then water does not need to be released if the river is no higher than 16.0 ft. at Austin and 18.0 ft. at Bastrop. The second rule is: if the predicted level of Lake Travis is between 683 and 685 feet msl then release up to 3,000 cfs if the river, with the release, is no higher than 16.0 ft. at Austin and 18.0 ft. at Bastrop. The third rule is: if the predicted level of Lake Travis is between 685 and 691 feet msl then release up to 5,000 cfs if the river, with the release, is no higher than 20.5 ft. at Austin and 25.1 ft. at Bastrop. The previous three new extracted rules are useful where they refine the original coarse rule based on the combination of the upstream (Lake Travis) and downstream (Lake Austin and Bastrop) conditions which primarily determine the decision on releasing water. More comprehensive rules can be extracted by applying the Full-RE technique.

11   Application of the Statistical Approach

The statistical approaches introduced in Section 3 were not used in the water reservoirs control problem because the extracted DORs were reasonably comprehensive. In this section, we report results of applying the statistical module to the breast cancer classification problem where there is no pre-existing domain knowledge (i.e., no DORs) but a public domain

Table 2. Rules extracted from network *"Cancer-Bin"* by BIO-RE technique.

No.	Rule Body	Cancer Class	Performance Measures		
			Sound-ness	Complete-ness	False Alarm
R_1	If $X_3 \leq 3.0$ and $X_7 \leq 3.3$ and $X_8 \leq 2.7$ and $X_9 \leq 1.5$	Benign	391/444	391/444	2/683
R_2	If $X_1 \leq 4.1$ and $X_3 \leq 3.0$ and $X_7 \leq 3.3$ and $X_9 \leq 1.5$	Benign	317/444	8/444	0/683
R_3	If $X_1 \leq 4.1$ and $X_3 \leq 3.0$ and $X_8 \leq 2.7$ and $X_9 \leq 1.5$	Benign	316/444	7/444	0/683
R_4	If $X_1 \leq 4.1$ and $X_3 \leq 3.0$ and $X_7 \leq 3.3$ and $X_8 \leq 2.7$	Benign	316/444	7/444	0/683
R_5	If $X_1 \leq 4.1$ and $X_7 \leq 3.3$ and $X_8 \leq 2.7$ and $X_9 \leq 1.5$	Benign	314/444	5/444	0/683
R_6	If $X_1 \geq 4.1$ and $X_3 \geq 3.0$	Malignant	200/239	199/239	15/683
R_7	If $X_3 \geq 3.0$ and $X_7 \geq 3.3$	Malignant	187/239	27/239	2/683
R_8	If $X_3 \geq 3.0$ and $X_8 \geq 2.7$	Malignant	187/239	3/239	0/683
R_9	If $X_1 \geq 4.1$ and $X_7 \geq 3.3$	Malignant	167/239	7/239	1/683
R_{10}	If $X_1 \geq 4.1$ and $X_9 \geq 1.5$	Malignant	100/239	1	3
R_{11}	Default Class	Benign	5/444	5/444	0/239
Total For Benign Rules				423/444	2/683
Total For Malignant Rules				237/239	21/683
Overall Performance%				96.63%	3.37%

data set is available [3], [23].

The Wisconsin breast cancer data set has nine inputs $(X_1 \cdots X_9)$ and two output classes (*Benign* or *Malignant*). The available 683 instances were divided randomly into a training set of size 341 and a test set of size 342. In all experiments, an MLP network is trained using the backprop-agation algorithm with momentum as well as a regularization term [11]. The dimensionality of the breast-cancer input space is reduced from 9 to 6 inputs using PCA. BIO-RE, Partial-RE, and Full-RE are used to extract rules from Cancer-Bin, Cancer-Norm, and Cancer-Cont networks respec-tively, where the first network is trained with a binarized version of the available data, Cancer-Norm is trained with normalized input patterns, and Cancer-Cont is trained with the original data set after dimensionality reduction. Tables 2, 3, and 4 present three sets of ordered rules extracted by the three rule extraction techniques, along with the corresponding per-formance measures.

Table 3. Rules extracted from network *"Cancer-Norm"* by Partial-RE technique.

No.	Rule Body	Cancer Class	CF	Performance Measures		
				Sound-ness	Complete-ness	False Alarm
R_1	If $X_2 \leq 3.0$ and $X_3 \leq 3.0$ and $X_7 \leq 3.3$	Benign	0.99	412/444	412/444	6/683
R_2	If $X_1 \leq 4.1$ and $X_3 \leq 3.0$ and $X_7 \leq 3.3$	Benign	0.99	324/444	1/444	0/683
R_3	If $X_2 \geq 3.0$ and $X_3 \geq 3.0$	Malignant	0.99	222/239	219/239	15/683
R_4	If $X_1 \geq 4.1$ and $X_7 \leq 3.3$ and $X_8 \geq 2.7$	Malignant	0.99	137/239	8/239	0/683
R_5	If $X_1 \leq 4.1$ and $X_2 \leq 3.0$ and $X_7 \leq 3.3$	Benign	0.84	327/444	4/444	0/683
R_6	If $X_1 \geq 4.1$ and $X_2 \geq 3.0$	Malignant	0.99	198/239	2/239	0/683
R_7	If $X_1 \leq 4.1$ and $X_2 \leq 3.0$ and $X_3 \leq 3.0$	Benign	0.84	333/444	9/444	1/683
R_8	If $X_1 \geq 4.1$ and $X_3 \geq 3.0$	Malignant	0.99	200/239	3/239	2/683
R_9	If $X_2 \leq 3.0$ and $X_3 \leq 3.0$ and $X_8 \leq 2.7$	Benign	0.99	409/444	1/444	0/683
Total For Benign Rules					427/444	7/683
Total For Malignant Rules					232/239	17/683
Overall Performance%					96.49%	3.51%

Table 5 provides an overall comparison between the performance of the extracted rules and their corresponding trained networks. It shows that the three techniques were successfully used with approximately the same performance regardless of the nature of the training and testing data sets used for each network. Also, it shows that binarizing and scaling the breast cancer data set did not degrade the performance of the trained networks or of the rules extracted by BIO-RE and Partial-RE from these networks (*"Cancer-Bin" and "Cancer-Norm" respectively*). This is due to the fact that the original input features of the breast cancer problem have the same range (1,10). Table 6 shows the impact of the integration

Table 4. Rules extracted from network *"Cancer-Cont"* by Full-RE technique.

No.	Rule Body	Cancer Class	CF	Performance Measures		
				Sound-ness	Complete-ness	False Alarm
R_1	If $X_1 < 8$ and $X_3 < 3$	Benign	0.96	394/444	394/444	5/683
R_2	If $X_2 \geq 2$ and $X_7 \geq 3$	Malignant	0.83	227/239	223/239	18/683
R_3	If $X_1 < 8$ and $X_7 < 3$	Benign	0.75	300/444	27/444	1/683
R_4	If $X_1 \geq 8$	Malignant	0.89	123/239	9/239	1/683
R_5	If $X_1 < 8$ and $X_2 < 2$	Benign	0.79	369/444	4/444	1/683
Total For Benign Rules					425/444	7/683
Total For Malignant Rules					232/239	19/683
Overall Performance%					96.19%	3.80%

Table 5. Performance comparison between the sets of extracted rules and their corresponding trained networks for the breast-cancer problem.

		Neural Network		Extracted Rules	
		ratio	% correct	ratio	% correct
Binarized	Training	333/341	97.65	331/341	97.07
Network	Testing	317/342	92.69	329/342	96.20
(Cancer-Bin)	Overall	650/683	95.17	660/683	96.63
Normalized	Training	329/341	96.48	331/341	97.07
Network	Testing	325/342	95.03	328/342	95.91
(Cancer-Norm)	Overall	654/683	95.75	659/683	96.49
Continuous	Training	334/341	97.95	330/341	96.77
Network	Testing	331/342	96.78	327/342	95.61
(Cancer-Cont)	Overall	665/683	97.36	657/683	96.19

method. It is important to mention that the limited gains due to the integration is because of the high degree of agreement between the two modules. Only 22, 20, and 14 out of 683 outcomes were different respectively for the three experiments. The integration mechanism was able to select correctly 20, 17, and 14 of these mismatches respectively.

12   Discussion

Symbolic and neural subsystems can be combined in a wide variety of ways [24]. HIA has the flavor of a transformational hybrid system, whose prototype is the Knowledge Based Neural Network (KBNN) that

Table 6. Overall performance of HIA after applying the integration mechanism using the breast-cancer database.

	#both correct	#both wrong	#disagreed on	#correct decisions on mismatches
Cancer-Bin	647	14	22	20/22
Cancer-Norm	647	16	20	17/20
Cancer-Cont	653	16	14	14/14

	Overall Performance	
	ratio	% correct
Cancer-Bin	667/683	97.77
Cancer-Norm	664/683	97.22
Cancer-Cont	667/683	97.77

achieves theory refinement in four phases [8], [13], [15], [36], [42]:

- The rule base representation phase, where the initial domain knowledge is extracted and represented in a symbolic format (e.g., a rule base system).

- The mapping phase, where domain knowledge represented in symbolic form is mapped into an initial connectionist architecture.

- The learning phase, where this connectionist architecture is trained by a set of domain examples.

- The rule extraction phase, where the trained (and thus modified) connectionist architecture is mapped back to a rule based system to provide explanation power.

The main motivation of such KBNN systems is to incorporate the complementary features of knowledge based and neural network paradigms. Knowledge-Based Artificial Neural Network (KBANN) [41] is a notable system that maps domain knowledge, represented in propositional logic (Horn clauses), into a neural network architecture which is then trained using the backpropagation algorithm to refine its mapped domain knowledge. The KBANN algorithm has been applied to two problems from molecular biology and the reported results show that it generalizes better than other learning systems. However, KBANN maps binary rules into a neural network and it is not clear how it can deal with rules with certainty factors. Also, it adds new hidden nodes before the training phase starts, but the difficult question of *how many hidden nodes should be added* is not answered.

RAPTURE is another hybrid system for revising probabilistic knowledge bases that combines connectionist and symbolic learning methods [22]. RAPTURE is capable of dealing with rule based systems that use certainty factors. However, the structure of the mapped network by RAPTURE is domain dependent and the number of network layers, being determined by the hierarchy of the rule base, can become large. Another notable example is the Knowledge Based Conceptual Neural Network (KBCNN) model. KBCNN revises and learns knowledge on the basis of a neural network translated from a rule base which encodes the initial domain theory [8], [9]. In fact, the KBCNN model has some similarities with both KBANN and the RAPTURE.

Researchers have also combined connectionist systems with fuzzy logic to obtain Fuzzy Logic Neural Networks (FLNN). In FLNNs, the neural network subsystem is typically used to adapt membership functions of fuzzy variables [5], or to refine and extract fuzzy rules [20], [37], [38]. A standard fuzzy logic system has four components:

- A fuzzifier, which determines the degree of membership of a crisp input in a fuzzy set.

- A fuzzy rule base, which represents the fuzzy relationships between input-output fuzzy variables. The output of the fuzzy rule base is determined based on the degree of membership specified by the fuzzifier.

- An inference engine, which controls the rule base.

- A defuzzifier, which converts the output of the fuzzy rule base into a crisp value.

Such fuzzy logic systems often suffer from two main problems. First, designing the right membership function that represents each input and output variable may be nontrivial. A common approach is to design an initial membership function, usually triangular or trapezoidal in shape, and subsequently refine it using some heuristic. The second problem is the lack of a learning function that can be adapted to reason about the environment.

These two problems can be alleviated by combining both fuzzy logic and neural network paradigms. In FLNN hybrid systems, a neural network architecture is used to replace the fuzzy rule base module of standard fuzzy logic systems. The Max and Min functions are commonly used as activation functions in this network. Then, a supervised or unsupervised learning algorithm is used instead of the inference engine to adapt network parameters and/or architecture. After training the network with available domain examples, the adapted network is used to refine initial membership functions and fuzzy rule base. Moreover, it may be used to extract new fuzzy rules.

The first five modules of HIA are superficially similar to both KBNN and FLNN hybrid systems. HIA is capable of revising initial domain knowledge and extracting new rules based on training examples. However, it has the following distinguishing features: (1) it generates a uniform neural network architecture because of the constrained DOR format; (2) the neural network architecture generated by HIA has only three layers, independent of the initial rule base hierarchy; (3) it revises the input characterization parameters using a coarse coding fuzzification approach during the training phase which may enhance system performance; (4) it combines a statistical module along with its knowledge based and neural network modules to extract supplementary domain knowledge.

Much of the power of HIA is derived from its completeness. It provides a mechanism for conflict resolution among the extracted rules, and for optional integration of the refined expert system and the trained neural network.

For the problem of controlling the flood gates of the Colorado river in greater Austin, we observe that refining the input characterization along with the domain knowledge incorporated in a connectionist model substantially enhanced the generalization ability of that model. This application also showed the capability of HIA to extract new and useful operational rules from the trained connectionist module. The breast cancer classification problem shows how the statistical module of HIA can enhance the topology of the connectionist module in cases where there is no available prior domain knowledge or in cases where the input features have substantial redundancy. It is remarkable that, using the Full-RE technique, the data set can be characterized with high accuracy using only five rules.

It will be worthwhile to apply HIA to a wider range of problems where both domain knowledge and training data are available, but none is sufficiently comprehensive on its own.

Acknowledgments

This research was supported in part by ARO contracts DAAG55-98-1-0230 and DAAD19-99-1-0012, and NSF grant ECS-9900353. Ismail Taha was also supported by the Egyptian Government Graduate Fellowship in Electrical and Computer Engineering. We are thankful to Prof. B. Kuipers for bringing the Colorado river problem to our attention.

References

[1] Aggarwal, J.K., Ghosh, J., Nair, D., and Taha, I. (1996), "A comparative study of three paradigms for object recognition - bayesian statistics, neural networks and expert systems," *Advances In Image Understanding: A Festschrift for Azriel Rosenfeld*, Boyer, K. and Ahuja, N. (Eds.), IEEE Computer Society Press, pp. 241-262.

[2] Andrews, R., Diederich, J., and Tickle, A. (1995), "A survey and critique of techniques for extracting rules from trained artificial neural networks," *Knowledge-Based Systems*, vol. 8, no. 6, pp. 373-389.

[3] Bennett, K. and Mangasarian, O. (1992), "Robust linear programming discrimination of two linearly inseparable sets," *Optimization Methods and Software 1*, Gordon and Breach Science Publishers.

[4] Jain, L.C., and Martin, N.M. (Eds.) (1999), *Fusion of Neural Networks, Fuzzy Systems and Genetic Algorithms*, CRC Press.

[5] Challo, R., McLauchlan, R., Clark, D., and Omar, S. (1994), "A fuzzy neural hybrid system," *IEEE International Conference on Neural Networks*, Orlando, FL, vol. III, pp. 1654-1657.

[6] Fletcher, J. and Obradovic, Z. (1993), "Combining prior symbolic knowledge and constructive neural network learning," *Connection Science*, vol. 5, no. 3-4, pp. 365-375.

[7] Friedman, J.H. (1994), "An overview of predictive learning and function approximation," *From Statistics to Neural Networks, Proc. NATO/ASI Workshop*, Cherkassky, V., Friedman, J., and Wechsler, H., editors, Springer-Verlag, pp. 1-61.

[8] Fu, L. (1993), "Knowledge-based connectionism for revising domain theories," *IEEE Transactions on Systems, Man, and Cybernetics*, vol. 23, no. 1, pp. 173-182.

[9] Fu, L. (1994), *Neural Networks in Computer Intelligence*, McGraw-Hill, Inc.

[10] Gallant, S.I. (1988), "Connectionist expert systems," *Comm. of ACM*, vol. 31, no. 2, pp. 152-169.

[11] Ghosh, J. and Tumer, K. (1994), "Structural adaptation and generalization in supervised feed-forward networks," *Journal of Artificial Neural Networks*, vol. 1, no. 4, pp. 431-458.

[12] Giles, C. and Omlin, C. (1994), "Pruning recurrent neural networks for improved generalization performance," *IEEE Transactions on Neural Networks*, vol. 5, no. 5, pp. 848-851.

[13] Glover, C., Silliman, M., Walker, M., and Spelt, P. (1990), "Hybrid neural network and rule-based pattern recognition system capable of self-modification," *Proceedings of SPIE, Application of Artificial Intelligence VIII*, pp. 290-300.

[14] Gonzalez, R. (1993), *Digital Image Processing*, Addison-Wesley.

[15] Hendler, J. (1989), "Marker-passing over microfeatures: towards a hybrid symbolic/connectionist model," *Cognitive Science*, vol. 13, pp. 79-106.

[16] Jackson, P. (1990), *Introduction to Expert Systems*, Addison-Wesley.

[17] Kanal, L. (1974), "Patterns in pattern recognition," *IEEE Trans. Information Theory*, IT-20:697-722.

[18] Lacher, R., Hruska, S., and Kuncicky, D. (1992), "Backpropagation learning in expert networks," *IEEE Transactions on Neural Networks*, vol. 3, no. 1, pp. 62-72.

[19] LCRA (1992). "The flooding of the Colorado: how the system worked to protect central Texas," Technical report, Lower Colorado River Authority, P.O. Box 220, Austin, Texas 78767-0220.

[20] Lin, C. and Lee, C. (1991), "Neural-network-based fuzzy logic control and decision system," *IEEE Transactions on Computers*, vol. 40, no. 12, pp. 1320-1326.

[21] Liu, H. and Setiono, R. (1995), "Chi2: feature selection and discretization of numeric attributes," *Proceedings of the Seventh International Conference on Tools with Artificial Intelligence*, pp. 388-391.

[22] Mahoney, J. and Mooney, R. (1993), "Combining connectionist and symbolic learning to refine certainty factor rule bases," *Connection Science*, vol. 5, no. 3-4, pp. 339-364.

[23] Mangasarian, O. and Wolberg, H. (1990), "Cancer diagnosis via linear programming," *SIAM News*, vol. 23, pp. 5, pp. 1-18.

[24] McGarry, K., Wertmer, S., and MacIntyre, J. (1999), "Hybrid neural systems: from simple coupling to fully integrated neural networks," *Neural Computing Surveys*, vol. 2, pp. 62-93.

[25] Medsker, L.R. (1995), *Hybrid Intelligent Systems*, Kluwer Academic, Norwell, MA.

[26] Michalski, R. (1993), "Toward a unified theory of learning: multi-strategy task-adaptive learning," *Readings in Knowledge Acquisition and Learning: Automating the Construction and Improvement of Expert Systems*, Buchanan, B. and Wilkins, D., editors, Morgan Kaufmann, San Mateo.

[27] Minsky, M. (1991), "Logical versus analogical or symbolic versus connectionist or neat versus scruffy," *AI Magazine*, vol. 12, no. 2, pp. 34-51.

[28] Ourston, D. and Mooney, R. (1990), "Changing the rules: a comprehensive approach to theory refinement," *Proceedings of the Eighth National Conference on Artificial Intelligence*, AAAI Press, pp. 815-820.

[29] Ruddel, R. and Sangiovanni-Vincentelli, A. (1985), "Espresso-MV: algorithms for multiple-valued logic minimization," *Proceedings of Cust. Int. Circ. Conf.*, Portland.

[30] Scott, A., Clayton, J., and Gibson, E. (1991), *A Practical Guide to Knowledge Acquisition*, Addison-Wesley.

[31] Sharkey, A. (1996), "On combining artificial neural networks," *Connection Science*, vol. 8, no. 3/4, pp. 299-314.

[32] Sharkey, N. and Sharkey, A. (1994), "Understanding catastrophic interference in neural nets," Technical Report CS-94-4, Dept. of CS, Univ. of Sheffield, UK.

[33] Sun, R. (1994), *Integrating Rules and Connectionism for Robust Commonsense Reasoning*, John Wiley and Sons.

[34] Taha, I. (1997), *A Hybrid Intelligent Architecture for Revising Domain Knowledge*, Ph.D. thesis, Dept. of ECE, Univ. of Texas at Austin, December.

[35] Taha, I. and Ghosh, J. (1997), "Hybrid intelligent architecture and its application to water reservoir control," *International Journal of Smart Engineering Systems*, vol. 1, pp. 59-75.

[36] Taha, I. and Ghosh, J. (1999), "Symbolic interpretation of artificial neural networks," *IEEE Trans. Knowledge and Data Engineering*, vol. 11, no. 3, pp. 448-463.

[37] Takagi, H. and Hayashi, I. (1992), "NN-driven fuzzy reasoning," *Fuzzy Models for Pattern Recognition*, Bezdek, J. and Pal, S., editors, IEEE Press, pp. 496-512.

[38] Tazaki, E. and Inoue, N. (1994), "A generation methods for fuzzy rules using neural networks with planar lattice architecture," *IEEE International Conference on Neural Networks*, Orlando, FL, vol. III, pp. 1743-1748.

[39] Tickle, A.B., Andrews, R., Golea, M., and Diederich, J. (1998), "The truth will come to light: directions and challenges in extracting the knowledge embedded within trained artificial neural networks," *IEEE Transactions on Neural Networks*, vol. 9, no. 6, pp. 1057-1068.

[40] Towell, G. and Shavlik, J. (1993), "The extraction of refined rules from knowledge-based neural networks," *Machine Learning*, vol. 13, no. 1, pp. 71-101.

[41] Towell, G. and Shavlik, J. (1994), "Knowledge-based artificial neural networks," *Artificial Intelligence*, vol. 70, no. 1-2, pp. 119-165.

[42] Towell, G., Shavlik, J., and Noordwier, M. (1990), "Refinement of approximate domain theories by knowledge-based artificial neural network," *Proceedings of Eighth National Conference on Artificial Intelligence*, pp. 861-866.

[43] Wang, L. and Mendel, J. (1992), "Generating fuzzy rules by learning examples," *IEEE Transactions on Systems, Man, and Cybernetics*, vol. 22, no. 6, pp. 1414-1427.

[44] Wilson, A. and Hendler, J. (1993), "Linking symbolic and subsymbolic computing," *Connection Science*, vol. 5, no. 3-4, pp. 395-414.

CHAPTER 2

NEW RADIAL BASIS NEURAL NETWORKS AND THEIR APPLICATION IN A LARGE-SCALE HANDWRITTEN DIGIT RECOGNITION PROBLEM

N.B. Karayiannis
Department of Electrical and Computer Engineering
University of Houston
Houston, Texas 77204-4793
U.S.A.
Karayiannis@UH.EDU

S. Behnke
Institute of Computer Science
Free University of Berlin
Takustr. 9, 14195 Berlin
Germany
behnke@inf.fu-berlin.de

This chapter presents an axiomatic approach for reformulating radial basis function (RBF) neural networks. With this approach the construction of admissible RBF models is reduced to the selection of generator functions that satisfy certain properties. The selection of specific generator functions is based on criteria which relate to their behavior when the training of reformulated RBF networks is performed by gradient descent. This chapter also presents batch and sequential learning algorithms developed for reformulated RBF networks using gradient descent. These algorithms are used to train reformulated RBF networks to recognize handwritten digits from the NIST databases.

0-8493-2268-5/2000/$0.00+$.50

1   Introduction

A *radial basis function* (RBF) neural network is usually trained to map a vector $\mathbf{x}_k \in \mathbf{R}^{n_i}$ into a vector $\mathbf{y}_k \in \mathbb{R}^{n_o}$, where the pairs $(\mathbf{x}_k, \mathbf{y}_k), 1 \le k \le M$, form the *training set*. If this mapping is viewed as a function in the input space $\mathbf{R}^{n_i}$, learning can be seen as a function approximation problem. From this point of view, learning is equivalent to finding a surface in a multidimensional space that provides the best fit for the training data. Generalization is therefore synonymous with interpolation between the data points along the constrained surface generated by the fitting procedure as the optimum approximation to this mapping.

Broomhead and Lowe [3] were the first to explore the use of radial basis functions in the design of neural networks and to show how RBF neural networks model nonlinear relationships and implement interpolation. Micchelli [33] showed that RBF neural networks can produce an interpolating surface which exactly passes through all the pairs of the training set. However, the exact fit is neither useful nor desirable in practice as it may produce anomalous interpolation surfaces. Poggio and Girosi [38] viewed the training of RBF neural networks as an ill-posed problem, in the sense that the information in the training data is not sufficient to uniquely reconstruct the mapping in regions where data are not available. From this point of view, learning is closely related to classical approximation techniques such as generalized splines and regularization theory. Park and Sandberg [36], [37] proved that RBF neural networks with one layer of radial basis functions are capable of universal approximation. Under certain mild conditions on radial basis functions, RBF networks are capable of approximating arbitrarily well any function. Similar proofs also exist in the literature for feed-forward neural models with sigmoidal nonlinearities [7].

The performance of a RBF neural network depends on the number and positions of the radial basis functions, their shapes, and the method used for learning the input-output mapping. The existing learning strategies for RBF neural networks can be classified as follows: 1) strategies selecting radial basis function centers randomly from the training data [3], 2) strategies employing unsupervised procedures for selecting radial basis function centers [5], [6], [25], [34], and 3) strategies employing supervised procedures for selecting radial basis function centers [4], [13], [17], [20], [21], [38].

Broomhead and Lowe [3] suggested that, in the absence of *a priori* knowledge, the centers of the radial basis functions can either be distributed uniformly within the region of the input space for which there is data, or chosen to be a subset of training points by analogy with strict interpolation. This approach is sensible only if the training data are distributed in a representative manner for the problem under consideration, an assumption that is very rarely satisfied in practical applications. Moody and Darken [34] proposed a hybrid learning process for training RBF neural networks with Gaussian radial basis functions, which is widely used in practice. This learning procedure employs different schemes for updating the *output weights*, i.e., the weights that connect the radial basis functions with the output units, and the centers of the radial basis functions, i.e., the vectors in the input space that represent the *prototypes* of the input vectors included in the training set. Moody and Darken used the *c*-means (or *k*-means) clustering algorithm [2] and a "*P*-nearest-neighbor" heuristic to determine the positions and widths of the Gaussian radial basis functions, respectively. The output weights are updated using a supervised least-mean-squares learning rule. Poggio and Girosi [38] proposed a fully supervised approach for training RBF neural networks with Gaussian radial basis functions, which updates the radial basis function centers together with the output weights. Poggio and Girosi used Green's formulas to deduct an optimal solution with respect to the objective function and employed gradient descent to approximate the regularized solution. They also proposed that Kohonen's self-organizing feature map [29], [30] can be used for initializing the radial basis function centers before gradient descent is used to adjust all of the free parameters of the network. Chen *et al.* [5], [6] proposed a learning procedure for RBF neural networks based on the *orthogonal least squares* (OLS) method. The OLS method is used as a forward regression procedure to select a suitable set of radial basis function centers. In fact, this approach selects radial basis function centers one by one until an adequate RBF neural network has been constructed. Cha and Kassam [4] proposed a stochastic gradient training algorithm for RBF neural networks with Gaussian radial basis functions. This algorithm uses gradient descent to update all free parameters of RBF neural networks, which include the radial basis function centers, the widths of the Gaussian radial basis functions, and the output weights. Whitehead and Choate [42] proposed an evolutionary training algorithm for RBF neural networks. In this approach, the centers of radial basis functions are governed by space-filling curves whose parameters evolve genetically. This encoding causes each group of co-determined basis functions to evolve in order to fit a region of the input space. Roy *et al.* [40] proposed a set of learning principles that led to a training algorithm for a network that contains "truncated" radial basis functions and other types of hidden units. This algorithm uses random clustering and linear programming to design and train the network with polynomial time complexity.

Despite the existence of a variety of learning schemes, RBF neural networks are frequently trained in practice using variations of the learning scheme proposed by Moody and Darken [34]. These hybrid learning schemes determine separately the prototypes that represent the radial basis function centers according to some *unsupervised* clustering or vector quantization algorithm and update the output weights by a *supervised* procedure to implement the desired input-output mapping. These approaches were developed as a natural reaction to the long training times typically associated with the training of traditional feed-forward neural networks using gradient descent [28]. In fact, these hybrid learning schemes achieve fast training of RBF neural networks as a result of the strategy they employ for learning the desired input-output mapping. However, the same strategy prevents the training set from participating in the formation of the radial basis function centers, with a negative impact on the performance of trained RBF neural networks [25]. This created a wrong impression about the actual capabilities of an otherwise powerful neural model. The training of RBF neural networks using gradient descent offers a solution to the trade-off between performance and training speed. Moreover, such training can make RBF neural networks serious competitors to feed-forward neural networks with sigmoidal hidden units.

Learning schemes attempting to train RBF neural networks by fixing the locations of the radial basis function centers are very slightly affected by the specific form of the radial basis functions used. On the other hand, the convergence of gradient descent learning and the performance of the trained RBF neural networks are both affected rather strongly by the choice of radial basis functions. The search for admissible radial basis functions other than the Gaussian function motivated the development of an axiomatic approach for constructing reformulated RBF neural networks suitable for gradient descent learning [13], [17], [20], [21].

2   Function Approximation Models and RBF Neural Networks

There are many similarities between RBF neural networks and function approximation models used to perform interpolation. Such function approximation models attempt to determine a surface in a Euclidean space $\mathbb{R}^{n_i}$ that provides the best fit for the data $(\mathbf{x}_k, y_k)$, $1 \le k \le M$, where $\mathbf{x}_k \in \mathcal{X} \subset \mathbb{R}^{n_i}$ and $y_k \in \mathbb{R}$ for all $k = 1, 2, \ldots M$. Micchelli [33] considered the solution of the interpolation problem $s(\mathbf{x}_k) = y_k, 1 \le k \le M$, by functions $s : \mathbb{R}^{n_i} \to \mathbb{R}$ of the form:

$$s(\mathbf{x}) = \sum_{k=1}^{M} w_k \, g(\|\mathbf{x} - \mathbf{x}_k\|^2). \tag{1}$$

This formulation treats interpolation as a function approximation problem, with the function $s(\cdot)$ generated by the fitting procedure as the best approximation to this mapping. Given the form of the basis function $g(\cdot)$, the function approximation problem described by $s(\mathbf{x}_k) = y_k, 1 \le k \le M$, reduces to determining the weights $w_k, 1 \le k \le M$, associated with the model (1).

The model described by equation (1) is admissible for interpolation if the basis function $g(\cdot)$ satisfies certain conditions. Micchelli [33] showed that a function $g(\cdot)$ can be used to solve this interpolation problem if the $M \times M$ matrix $\mathbf{G} = [g_{ij}]$ with entries $g_{ij} = g(\|\mathbf{x}_i - \mathbf{x}_j\|^2)$ is positive definite. The matrix $\mathbf{G}$ is positive definite if the function $g(\cdot)$ is *completely monotonic* on $(0, \infty)$. A function $g(\cdot)$ is called completely monotonic on $(0, \infty)$ if it is continuous on $(0, \infty)$ and its ℓth order derivatives $g^{(\ell)}(x)$ satisfy $(-1)^\ell g^{(\ell)}(x) \ge 0$, $\forall x \in (0, \infty)$, for $\ell = 0, 1, 2, \ldots$.

RBF neural network models can be viewed as the natural extension of this formulation. Consider the function approximation model described by:

$$\hat{y} = w_0 + \sum_{j=1}^{c} w_j \, g(\|\mathbf{x} - \mathbf{v}_j\|^2). \tag{2}$$

If the function $g(\cdot)$ satisfies certain conditions, the model (2) can be used to implement a desired mapping $\mathbb{R}^{n_i} \to \mathbb{R}$ specified by the training set $(\mathbf{x}_k, y_k)$, $1 \le k \le M$. This is usually accomplished by devising a learning procedure to determine its adjustable parameters. In addition to the weights w_j, $0 \le j \le c$, the adjustable parameters of the model (2) also include the vectors $\mathbf{v}_j \in \mathcal{V} \subset \mathbb{R}^{n_i}$, $1 \le j \le c$. These vectors are determined during learning as the prototypes of the input vectors $\mathbf{x}_k$, $1 \le k \le M$. The adjustable parameters of the model (2) are frequently updated by minimizing some measure of the discrepancy between the expected output y_k of the model to the corresponding input $\mathbf{x}_k$ and its actual response:

$$\hat{y}_k = w_0 + \sum_{j=1}^{c} w_j \, g(\|\mathbf{x}_k - \mathbf{v}_j\|^2), \tag{3}$$

for all pairs $(\mathbf{x}_k, y_k)$, $1 \le k \le M$, included in the training set.

The function approximation model (2) can be extended to implement any mapping $\mathbb{R}^{n_i} \to \mathbb{R}^{n_o}$, $n_o \ge 1$, as:

$$\hat{y}_i = f\left(w_{i0} + \sum_{j=1}^{c} w_{ij} \, g(\|\mathbf{x} - \mathbf{v}_j\|^2) \right), 1 \le i \le n_o, \tag{4}$$

where $f(\cdot)$ is a non-decreasing, continuous and differentiable everywhere function. The model (4) describes a RBF neural network with inputs from $\mathbb{R}^{n_i}$, c radial basis function units, and n_o output units if:

$$g(x^2) = \phi(x), \tag{5}$$

and $\phi(\cdot)$ is a radial basis function. In such a case, the response of the network to the input vector $\mathbf{x}_k$ is:

$$\hat{y}_{i,k} = f\left(\sum_{j=0}^{c} w_{ij} \, h_{j,k} \right), 1 \le i \le n_o, \tag{6}$$

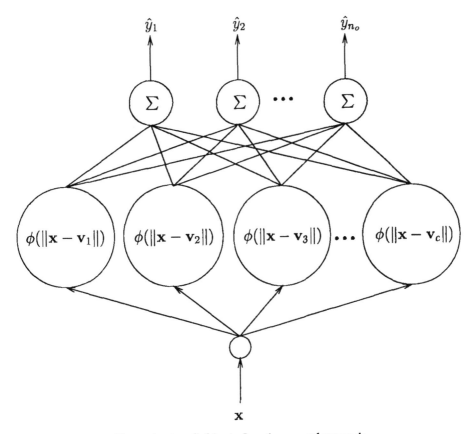

Figure 1. A radial basis function neural network.

where $h_{0,k} = 1$, $\forall k$, and $h_{j,k}$ represents the response of the radial basis function located at the jth prototype to the input vector $\mathbf{x}_k$, that is,

$$
\begin{aligned}
h_{j,k} &= \phi(\|\mathbf{x}_k - \mathbf{v}_j\|) \\
&= g(\|\mathbf{x}_k - \mathbf{v}_j\|^2), \; 1 \le j \le c.
\end{aligned}
\tag{7}
$$

The response (6) of the RBF neural network to the input $\mathbf{x}_k$ is actually the output of the upper associative network. When the RBF neural network is presented with $\mathbf{x}_k$, the input of the upper associative network is formed by the responses (7) of the radial basis functions located at the prototypes $\mathbf{v}_j, 1 \le j \le c$, as shown in Figure 1.

The models used in practice to implement RBF neural networks usually contain linear output units. An RBF model with linear output units can be seen as a special case of the model (4) that corresponds to $f(x) = x$. The choice of linear output units was mainly motivated by the hybrid learning schemes originally developed for training RBF neural networks. Nevertheless, the learning process is only slightly affected by the form of $f(\cdot)$ if RBF neural networks are trained using learning algorithms based on gradient descent. Moreover, the form of an admissible function $f(\cdot)$ does not affect the function approximation capability of the model (4) or the conditions that must be satisfied by radial basis functions. Finally, the use of a nonlinear sigmoidal function $f(\cdot)$ could make RBF models stronger competitors to traditional feed-forward neural networks in certain applications, such as those involving pattern classification.

3   Reformulating Radial Basis Neural Networks

A RBF neural network is often interpreted as a composition of localized receptive fields. The locations of these receptive fields are determined by the prototypes while their shapes are determined by the radial basis functions used. The interpretation often associated with RBF neural networks imposes some implicit constraints on the selection of radial basis functions. For example, RBF neural networks often employ decreasing Gaussian radial basis functions despite the fact that there exist both increasing and decreasing radial basis functions. The "neural" interpretation of the model (4) can be the basis of a systematic search for radial basis functions to be used for reformulating RBF neural networks [13], [17], [20], [21]. Such a systematic search is based on mathematical restrictions imposed on radial basis functions by their role in the formation of receptive fields.

The interpretation of a RBF neural network as a composition of receptive fields requires that the responses of all radial basis functions to all inputs are always positive. If the prototypes are interpreted as the centers of receptive fields, it is required that the response of any radial basis function becomes stronger as the input approaches its corresponding prototype. Finally, it is required that the response of any radial basis function becomes more sensitive to an input vector as this input vector approaches its corresponding prototype.

Let $h_{j,k} = g\left(\|\mathbf{x}_k - \mathbf{v}_j\|^2\right)$ be the response of the jth radial basis function of a RBF neural network to the input $\mathbf{x}_k$. According to the above interpretation of RBF neural networks, any admissible radial basis function $\phi(x) = g(x^2)$ must satisfy the following three axiomatic requirements [13], [17], [20], [21]:

Axiom 1: $h_{j,k} > 0$ for all $\mathbf{x}_k \in \mathcal{X}$ and $\mathbf{v}_j \in \mathcal{V}$.

Axiom 2: $h_{j,k} > h_{j,\ell}$ for all $\mathbf{x}_k, \mathbf{x}_\ell \in \mathcal{X}$ and $\mathbf{v}_j \in \mathcal{V}$ such that $\|\mathbf{x}_k - \mathbf{v}_j\|^2 < \|\mathbf{x}_\ell - \mathbf{v}_j\|^2$.

Axiom 3: If $\nabla_{\mathbf{x}_k} h_{j,k} \equiv \partial h_{j,k}/\partial \mathbf{x}_k$ denotes the gradient of $h_{j,k}$ with respect to the corresponding input $\mathbf{x}_k$, then:

$$\frac{\|\nabla_{\mathbf{x}_k} h_{j,k}\|^2}{\|\mathbf{x}_k - \mathbf{v}_j\|^2} > \frac{\|\nabla_{\mathbf{x}_\ell} h_{j,\ell}\|^2}{\|\mathbf{x}_\ell - \mathbf{v}_j\|^2},$$

for all $\mathbf{x}_k, \mathbf{x}_\ell \in \mathcal{X}$ and $\mathbf{v}_j \in \mathcal{V}$ such that $\|\mathbf{x}_k - \mathbf{v}_j\|^2 < \|\mathbf{x}_\ell - \mathbf{v}_j\|^2$.

These basic axiomatic requirements impose some rather mild mathematical restrictions on the search for admissible radial basis functions. Nevertheless, this search can be further restricted by imposing additional requirements that lead to stronger mathematical conditions. For example, it is reasonable to require that the responses of all radial basis functions to all inputs are bounded, i.e., $h_{j,k} < \infty, \forall j, k$. On the other hand, the third axiomatic requirement can be made stronger by requiring that:

$$\|\nabla_{\mathbf{x}_k} h_{j,k}\|^2 > \|\nabla_{\mathbf{x}_\ell} h_{j,\ell}\|^2 \tag{8}$$

if $\|\mathbf{x}_k - \mathbf{v}_j\|^2 < \|\mathbf{x}_\ell - \mathbf{v}_j\|^2$. Since $\|\mathbf{x}_k - \mathbf{v}_j\|^2 < \|\mathbf{x}_\ell - \mathbf{v}_j\|^2$,

$$\frac{\|\nabla_{\mathbf{x}_k} h_{j,k}\|^2}{\|\mathbf{x}_k - \mathbf{v}_j\|^2} > \frac{\|\nabla_{\mathbf{x}_k} h_{j,k}\|^2}{\|\mathbf{x}_\ell - \mathbf{v}_j\|^2}. \tag{9}$$

If $\|\nabla_{\mathbf{x}_k} h_{j,k}\|^2 > \|\nabla_{\mathbf{x}_\ell} h_{j,\ell}\|^2$ and $\|\mathbf{x}_k - \mathbf{v}_j\|^2 < \|\mathbf{x}_\ell - \mathbf{v}_j\|^2$, then:

$$\frac{\|\nabla_{\mathbf{x}_k} h_{j,k}\|^2}{\|\mathbf{x}_k - \mathbf{v}_j\|^2} > \frac{\|\nabla_{\mathbf{x}_k} h_{j,k}\|^2}{\|\mathbf{x}_\ell - \mathbf{v}_j\|^2} > \frac{\|\nabla_{\mathbf{x}_\ell} h_{j,\ell}\|^2}{\|\mathbf{x}_\ell - \mathbf{v}_j\|^2}, \tag{10}$$

and the third axiomatic requirement is satisfied. This implies that condition (8) is stronger than that imposed by the third axiomatic requirement.

The above discussion suggests two complementary axiomatic requirements for radial basis functions [17]:

Axiom 4: $h_{j,k} < \infty$ for all $x_k \in \mathcal{X}$ and $v_j \in \mathcal{V}$.

Axiom 5: If $\nabla_{x_k} h_{j,k} \equiv \partial h_{j,k} / \partial x_k$ denotes the gradient of $h_{j,k}$ with respect to the corresponding input x_k, then:

$$\|\nabla_{x_k} h_{j,k}\|^2 > \|\nabla_{x_\ell} h_{j,\ell}\|^2,$$

for all $x_k, x_\ell \in \mathcal{X}$ and $v_j \in \mathcal{V}$ such that $\|x_k - v_j\|^2 < \|x_\ell - v_j\|^2$.

The selection of admissible radial basis functions can be facilitated by the following theorem [17]:

Theorem 1: The model described by equation (4) represents a RBF neural network in accordance with all five axiomatic requirements if and only if $g(\cdot)$ is a continuous function on $(0, \infty)$, such that:

1. $g(x) > 0, \forall x \in (0, \infty)$,

2. $g(x)$ is a monotonically decreasing function of $x \in (0, \infty)$, i.e., $g'(x) < 0, \forall x \in (0, \infty)$,

3. $g'(x)$ is a monotonically increasing function of $x \in (0, \infty)$, i.e., $g''(x) > 0, \forall x \in (0, \infty)$,

4. $\lim_{x \to 0_+} g(x) = L$, where L is a finite number.

5. $d(x) = g'(x) + 2x\, g''(x) > 0, \forall x \in (0, \infty)$.

A radial basis function is said to be *admissible in the wide sense* if it satisfies the three basic axiomatic requirements, that is, the first three conditions of Theorem 1 [13], [17], [20], [21]. If a radial basis function satisfies all five axiomatic requirements, that is, all five conditions of Theorem 1, then it is said to be *admissible in the strict sense* [17].

A systematic search for admissible radial basis functions can be facilitated by considering basis functions of the form $\phi(x) = g(x^2)$, with $g(\cdot)$
defined in terms of a *generator function* $g_0(\cdot)$ as $g(x) = (g_0(x))^{\frac{1}{1-m}}$,
$m \neq 1$ [13], [17], [20], [21]. The selection of generator functions that
lead to admissible radial basis functions can be facilitated by the following theorem [17]:

Theorem 2: Consider the model (4) and let $g(x)$ be defined in terms of
the generator function $g_0(x)$ that is continuous on $(0, \infty)$ as:

$$g(x) = (g_0(x))^{\frac{1}{1-m}}, m \neq 1. \tag{11}$$

If $m > 1$, then this model represents a RBF neural network in accordance
with all five axiomatic requirements if:

1. $g_0(x) > 0$, $\forall x \in (0, \infty)$,

2. $g_0(x)$ is a monotonically increasing function of $x \in (0, \infty)$, i.e.,
   $g_0'(x) > 0, \forall x \in (0, \infty)$,

3. $r_0(x) = \frac{m}{m-1}(g_0'(x))^2 - g_0(x) g_0''(x) > 0$, $\forall x \in (0, \infty)$,

4. $\lim_{x \to 0_+} g_0(x) = L_1 > 0$,

5. $d_0(x) = g_0(x) g_0'(x) - 2 x r_0(x) < 0$, $\forall x \in (0, \infty)$.

If $m < 1$, then this model represents a RBF neural network in accordance
with all five axiomatic requirements if:

1. $g_0(x) > 0$, $\forall x \in (0, \infty)$,

2. $g_0(x)$ is a monotonically decreasing function of $x \in (0, \infty)$, i.e.,
   $g_0'(x) < 0, \forall x \in (0, \infty)$,

3. $r_0(x) = \frac{m}{m-1}(g_0'(x))^2 - g_0(x) g_0''(x) < 0$, $\forall x \in (0, \infty)$,

4. $\lim_{x \to 0_+} g_0(x) = L_2 < \infty$,

5. $d_0(x) = g_0(x) g_0'(x) - 2 x r_0(x) > 0$, $\forall x \in (0, \infty)$.

Any generator function that satisfies the first three conditions of Theorem 2 leads to admissible radial basis functions in the wide sense [13], [17], [20], [21]. Admissible radial basis functions in the strict sense can be obtained from generator functions that satisfy all five conditions of Theorem 2 [17].

4   Admissible Generator Functions

This section investigates the admissibility in the wide and strict sense of linear and exponential generator functions.

4.1   Linear Generator Functions

Consider the function $g(x) = (g_0(x))^{\frac{1}{1-m}}$, with $g_0(x) = a\,x + b$ and $m > 1$. Clearly, $g_0(x) = a\,x + b > 0, \forall x \in (0, \infty)$, for all $a > 0$ and $b \geq 0$. Moreover, $g_0(x) = a\,x + b$ is a monotonically increasing function if $g_0'(x) = a > 0$. For $g_0(x) = a\,x + b$, $g_0'(x) = a$, $g_0''(x) = 0$, and

$$r_0(x) = \frac{m}{m-1}\,a^2. \tag{12}$$

If $m > 1$, then $r_0(x) > 0, \forall x \in (0, \infty)$. Thus, $g_0(x) = a\,x + b$ is an admissible generator function in the wide sense (i.e., in the sense that it satisfies the three basic axiomatic requirements) for all $a > 0$ and $b \geq 0$. Certainly, all combinations of $a > 0$ and $b > 0$ also lead to admissible generator functions in the wide sense.

For $g_0(x) = a\,x + b$, the fourth axiomatic requirement is satisfied if:

$$\lim_{x \to 0+} g_0(x) = b > 0. \tag{13}$$

For $g_0(x) = a\,x + b$,

$$d_0(x) = (a\,x + b)\,a - 2\,x\,\frac{m}{m-1}\,a^2. \tag{14}$$

If $m > 1$, the fifth axiomatic requirement is satisfied if $d_0(x) < 0, \forall x \in (0, \infty)$. For $a > 0$, the condition $d_0(x) < 0$ is satisfied by $g_0(x) = a\,x + b$ if:

$$x > \frac{m-1}{m+1}\,\frac{b}{a}. \tag{15}$$

Since $m > 1$, the fifth axiomatic requirement is satisfied only if $b = 0$ or, equivalently, if $g_0(x) = a\,x$. However, the value $b = 0$ violates the fourth axiomatic requirement. Thus, there exists no combination of $a > 0$ and $b > 0$ leading to an admissible generator function in the strict sense that has the form $g_0(x) = a\,x + b$.

If $a = 1$ and $b = \gamma^2$, then the linear generator function $g_0(x) = a\,x + b$ becomes $g_0(x) = x + \gamma^2$. For this generator function, $g(x) = (x + \gamma^2)^{\frac{1}{1-m}}$. If $m = 3$, $g(x) = (x + \gamma^2)^{-\frac{1}{2}}$ corresponds to the inverse multiquadratic radial basis function:

$$\phi(x) = g(x^2) = \frac{1}{(x^2 + \gamma^2)^{\frac{1}{2}}}. \tag{16}$$

For $g_0(x) = x + \gamma^2$, $\lim_{x \to 0_+} g_0(x) = \gamma^2$ and $\lim_{x \to 0_+} g(x) = \gamma^{\frac{2}{1-m}}$. Since $m > 1$, $g(\cdot)$ is a bounded function if γ takes nonzero values. However, the bound of $g(\cdot)$ increases and approaches infinity as γ decreases and approaches 0. If $m > 1$, the condition $d_0(x) < 0$ is satisfied by $g_0(x) = x + \gamma^2$ if:

$$x > \frac{m-1}{m+1}\gamma^2. \tag{17}$$

Clearly, the fifth axiomatic requirement is satisfied only for $\gamma = 0$, which leads to an unbounded function $g(\cdot)$ [13], [20], [21].

Another useful generator function for practical applications can be obtained from $g_0(x) = a\,x + b$ by selecting $b = 1$ and $a = \delta > 0$. For $g_0(x) = 1 + \delta\,x$, $\lim_{x \to 0_+} g(x) = \lim_{x \to 0_+} g_0(x) = 1$. For this choice of parameters, the corresponding radial basis function $\phi(x) = g(x^2)$ is bounded by 1, which is also the bound of the Gaussian radial basis function. If $m > 1$, the condition $d_0(x) < 0$ is satisfied by $g_0(x) = 1 + \delta\,x$ if:

$$x > \frac{m-1}{m+1}\frac{1}{\delta}. \tag{18}$$

For a fixed $m > 1$, the fifth axiomatic requirement is satisfied in the limit $\delta \to \infty$. Thus, a reasonable choice for δ in practical situations is $\delta \gg 1$.

The radial basis function that corresponds to the linear generator function $g_0(x) = a\,x + b$ and some value of $m > 1$ can also be obtained from the decreasing function $g_0(x) = 1/(a\,x + b)$ combined with an appropriate value of $m < 1$. As an example, for $m = 3$, the generator function $g_0(x) = a\,x + b$ leads to $g(x) = (a\,x + b)^{-\frac{1}{2}}$. For $a = 1$ and $b = \gamma^2$, this generator function corresponds to the multiquadratic radial basis function (16). The multiquadratic radial basis function (16) can also be obtained using the decreasing generator function $g_0(x) = 1/(x + \gamma^2)$ with $m = -1$. In general, the function $g(x) = (g_0(x))^{\frac{1}{1-m}}$ corresponding to the increasing generator function $g_0(x) = a\,x + b$ and $m = m_i > 1$, is identical with the function $g(x) = (g_0(x))^{\frac{1}{1-m}}$ corresponding to the decreasing function $g_0(x) = 1/(a\,x + b)$ and $m = m_d$ if:

$$\frac{1}{1 - m_i} = \frac{1}{m_d - 1}, \tag{19}$$

or, equivalently, if:

$$m_i + m_d = 2. \tag{20}$$

Since $m_i > 1$, (20) implies that $m_d < 1$.

The admissibility of the decreasing generator function $g_0(x) = 1/(a\,x + b)$ can also be verified by using directly the results of Theorem 2. Consider the function $g(x) = (g_0(x))^{\frac{1}{1-m}}$, with $g_0(x) = 1/(a\,x + b)$ and $m < 1$. For all $a > 0$ and $b > 0$, $g_0(x) = 1/(a\,x + b) > 0$, $\forall x \in (0, \infty)$. Since $g_0'(x) = -a/(a\,x + b)^2 < 0$, $\forall x \in (0, \infty)$, $g_0(x) = 1/(a\,x + b)$ is a monotonically decreasing function for all $a > 0$. Since $g_0''(x) = 2a^2/(a\,x + b)^3$,

$$r_0(x) = \frac{2 - m}{m - 1} \frac{a^2}{(a\,x + b)^4}. \tag{21}$$

For $m < 1$, $r_0(x) < 0$, $\forall x \in (0, \infty)$, and $g_0(x) = 1/(a\,x + b)$ is an admissible generator function in the wide sense.

For $g_0(x) = 1/(a\,x + b)$,

$$\lim_{x \to 0_+} g_0(x) = \frac{1}{b}, \tag{22}$$

which implies that $g_0(x) = 1/(a\,x + b)$ satisfies the fourth axiomatic requirement unless b approaches 0. In such a case, $\lim_{x \to 0_+} g_0(x) = 1/b =$

∞. For $g_0(x) = 1/(a\,x + b)$,

$$d_0(x) = \frac{a}{(a\,x + b)^4}\left(a\,\frac{m-3}{m-1}x - b\right). \tag{23}$$

If $m < 1$, the fifth axiomatic requirement is satisfied if $d_0(x) > 0, \forall x \in (0, \infty)$. Since $a > 0$, the condition $d_0(x) > 0$ is satisfied by $g_0(x) = 1/(a\,x + b)$ if:

$$x > \frac{m-1}{m-3}\frac{b}{a}. \tag{24}$$

Once again, the fifth axiomatic requirement is satisfied for $b = 0$, a value that violates the fourth axiomatic requirement.

4.2   Exponential Generator Functions

Consider the function $g(x) = (g_0(x))^{\frac{1}{1-m}}$, with $g_0(x) = \exp(\beta x)$, $\beta > 0$, and $m > 1$. For any β, $g_0(x) = \exp(\beta x) > 0, \forall x \in (0, \infty)$. For all $\beta > 0$, $g_0(x) = \exp(\beta x)$ is a monotonically increasing function of $x \in (0, \infty)$. For $g_0(x) = \exp(\beta x)$, $g_0'(x) = \beta \exp(\beta x)$ and $g_0''(x) = \beta^2 \exp(\beta x)$. In this case,

$$r_0(x) = \frac{1}{m-1}(\beta \exp(\beta x))^2. \tag{25}$$

If $m > 1$, then $r_0(x) > 0, \forall x \in (0, \infty)$. Thus, $g_0(x) = \exp(\beta x)$ is an admissible generator function in the wide sense for all $\beta > 0$.

For $g_0(x) = \exp(\beta x), \beta > 0$,

$$\lim_{x \to 0_+} g_0(x) = 1 > 0, \tag{26}$$

which implies that $g_0(x) = \exp(\beta x)$ satisfies the fourth axiomatic requirement. For $g_0(x) = \exp(\beta x), \beta > 0$,

$$d_0(x) = (\beta \exp(\beta x))^2 \left(\frac{1}{\beta} - \frac{2}{m-1}x\right). \tag{27}$$

For $m > 1$, the fifth axiomatic requirement is satisfied if $d_0(x) < 0$, $\forall x \in (0, \infty)$. The condition $d_0(x) < 0$ is satisfied by $g_0(x) = \exp(\beta x)$ if:

$$x > \frac{m-1}{2\beta} = \frac{\sigma^2}{2} > 0, \tag{28}$$

where $\sigma^2 = (m-1)/\beta$. Regardless of the value $\beta > 0$, $g_0(x) = \exp(\beta x)$ is not an admissible generator function in the strict sense.

Consider also the function $g(x) = (g_0(x))^{\frac{1}{1-m}}$, with $g_0(x) = \exp(-\beta x)$, $\beta > 0$, and $m < 1$. For any β, $g_0(x) = \exp(-\beta x) > 0$, $\forall x \in (0, \infty)$. For all $\beta > 0$, $g_0'(x) = -\beta \exp(-\beta x) < 0$, $\forall x \in (0, \infty)$, and $g_0(x) = \exp(-\beta x)$ is a monotonically decreasing function. Since $g_0''(x) = \beta^2 \exp(-\beta x)$,

$$r_0(x) = \frac{1}{m-1} \left(\beta \exp(-\beta x) \right)^2 . \tag{29}$$

If $m < 1$, then $r_0(x) < 0$, $\forall x \in (0, \infty)$, and $g_0(x) = \exp(-\beta x)$ is an admissible generator function in the wide sense for all $\beta > 0$.

For $g_0(x) = \exp(-\beta x)$, $\beta > 0$,

$$\lim_{x \to 0+} g_0(x) = 1 < \infty, \tag{30}$$

which implies that $g_0(x) = \exp(-\beta x)$ satisfies the fourth axiomatic requirement. For $g_0(x) = \exp(-\beta x)$, $\beta > 0$,

$$d_0(x) = (\beta \exp(-\beta x))^2 \left(-\frac{1}{\beta} + \frac{2}{1-m} x \right). \tag{31}$$

For $m < 1$, the fifth axiomatic requirement is satisfied if $d_0(x) > 0$, $\forall x \in (0, \infty)$. The condition $d_0(x) > 0$ is satisfied by $g_0(x) = \exp(-\beta x)$ if:

$$x > \frac{1-m}{2\beta} = \frac{\sigma^2}{2}, \tag{32}$$

where $\sigma^2 = (1-m)/\beta$. Once again, $g_0(x) = \exp(-\beta x)$ is not an admissible generator function in the strict sense regardless of the value of $\beta > 0$.

It must be emphasized that both increasing and decreasing exponential generator functions essentially lead to the same radial basis function. If $m > 1$, the increasing exponential generator function $g_0(x) = \exp(\beta x)$, $\beta > 0$, corresponds to the Gaussian radial basis function $\phi(x) = g(x^2) = \exp(-x^2/\sigma^2)$, with $\sigma^2 = (m-1)/\beta$. If $m < 1$, the decreasing exponential generator function $g_0(x) = \exp(-\beta x)$, $\beta > 0$, also corresponds to the Gaussian radial basis function $\phi(x) = g(x^2) = \exp(-x^2/\sigma^2)$, with $\sigma^2 = (1-m)/\beta$. In fact, the function $g(x) = (g_0(x))^{\frac{1}{1-m}}$ corresponding to the increasing generator function $g_0(x) = \exp(\beta x), \beta > 0$, with $m = m_i > 1$ is identical with the function $g(x) = (g_0(x))^{\frac{1}{1-m}}$ corresponding to the decreasing function $g_0(x) = \exp(-\beta x), \beta > 0$, with $m = m_d < 1$ if:

$$m_i - 1 = 1 - m_d, \tag{33}$$

or, equivalently, if:

$$m_i + m_d = 2. \tag{34}$$

5   Selecting Generator Functions

All possible generator functions considered in the previous section satisfy the three basic axiomatic requirements but none of them satisfies all five axiomatic requirements. In particular, the fifth axiomatic requirement is satisfied only by generator functions of the form $g_0(x) = a\,x$, which violate the fourth axiomatic requirement. Therefore, it is clear that at least one of the five axiomatic requirements must be compromised in order to select a generator function. Since the response of the radial basis functions must be bounded in some function approximation applications, generator functions can be selected by compromising the fifth axiomatic requirement. Although this requirement is by itself very restrictive, its implications can be used to guide the search for generator functions appropriate for gradient descent learning [17].

5.1   The Blind Spot

Since $h_{j,k} = g(\|\mathbf{x}_k - \mathbf{v}_j\|^2)$,

$$
\begin{aligned}
\nabla_{\mathbf{x}_k} h_{j,k} &= g'(\|\mathbf{x}_k - \mathbf{v}_j\|^2)\,\nabla_{\mathbf{x}_k}(\|\mathbf{x}_k - \mathbf{v}_j\|^2) \\
&= 2\,g'(\|\mathbf{x}_k - \mathbf{v}_j\|^2)\,(\mathbf{x}_k - \mathbf{v}_j).
\end{aligned}
\tag{35}
$$

The norm of the gradient $\nabla_{\mathbf{x}_k} h_{j,k}$ can be obtained from (35) as:

$$
\begin{aligned}
\|\nabla_{\mathbf{x}_k} h_{j,k}\|^2 &= 4\,\|\mathbf{x}_k - \mathbf{v}_j\|^2 \left(g'(\|\mathbf{x}_k - \mathbf{v}_j\|^2)\right)^2 \\
&= 4\,t(\|\mathbf{x}_k - \mathbf{v}_j\|^2),
\end{aligned}
\tag{36}
$$

where $t(x) = x\,(g'(x))^2$. According to Theorem 1, the fifth axiomatic requirement is satisfied if and only if $d(x) = g'(x) + 2\,x\,g''(x) > 0$, $\forall x \in (0, \infty)$. Since $t(x) = x\,(g'(x))^2$,

$$
\begin{aligned}
t'(x) &= g'(x)\,(g'(x) + 2\,x\,g''(x)) \\
&= g'(x)\,d(x).
\end{aligned}
\tag{37}
$$

Theorem 1 requires that $g(x)$ is a decreasing function of $x \in (0, \infty)$, which implies that $g'(x) < 0$, $\forall x \in (0, \infty)$. Thus, (37) indicates that the fifth axiomatic requirement is satisfied if $t'(x) < 0, \forall x \in (0, \infty)$. If this condition is not satisfied, then $\|\nabla_{\mathbf{x}_k} h_{j,k}\|^2$ is not a monotonically decreasing function of $\|\mathbf{x}_k - \mathbf{v}_j\|^2$ in the interval $(0, \infty)$, as required by the fifth axiomatic requirement. Given a function $g(\cdot)$ satisfying the three basic axiomatic requirements, the fifth axiomatic requirement can be relaxed by requiring that $\|\nabla_{\mathbf{x}_k} h_{j,k}\|^2$ is a monotonically decreasing function of $\|\mathbf{x}_k - \mathbf{v}_j\|^2$ in the interval (B, ∞) for some $B > 0$. According to (36), this is guaranteed if the function $t(x) = x\,(g'(x))^2$ has a maximum at $x = B$ or, equivalently, if there exists a $B > 0$ such that $t'(B) = 0$ and $t''(B) < 0$. If $B \in (0, \infty)$ is a solution of $t'(x) = 0$ and $t''(B) < 0$, then $t'(x) > 0, \forall x \in (0, B)$, and $t'(x) < 0, \forall x \in (B, \infty)$. Thus, $\|\nabla_{\mathbf{x}_k} h_{j,k}\|^2$ is an increasing function of $\|\mathbf{x}_k - \mathbf{v}_j\|^2$ for $\|\mathbf{x}_k - \mathbf{v}_j\|^2 \in (0, B)$ and a decreasing function of $\|\mathbf{x}_k - \mathbf{v}_j\|^2$ for $\|\mathbf{x}_k - \mathbf{v}_j\|^2 \in (B, \infty)$. For all input vectors $\mathbf{x}_k$ that satisfy $\|\mathbf{x}_k - \mathbf{v}_j\|^2 < B$, the norm of the gradient $\nabla_{\mathbf{x}_k} h_{j,k}$ corresponding to the jth radial basis function decreases as $\mathbf{x}_k$ approaches its center that is located at the prototype $\mathbf{v}_j$. This is exactly the opposite behavior of what would intuitively be expected, given the interpretation of radial basis functions as receptive fields. As far as gradient descent learning is concerned, the hypersphere $\mathcal{R}_B = \{\mathbf{x} \in \mathcal{X} \subset \mathbf{R}^{n_i} : \|\mathbf{x} - \mathbf{v}\|^2 \in (0, B)\}$ is a "blind spot" for the radial basis function located at the prototype $\mathbf{v}$. The blind spot provides a measure of the sensitivity of radial basis functions to input vectors close to their centers.

The blind spot $\mathcal{R}_{B_{\text{lin}}}$ corresponding to the linear generator function

$g_0(x) = a\,x + b$ is determined by:

$$B_{\text{lin}} = \frac{m-1}{m+1}\frac{b}{a}. \tag{38}$$

The effect of the parameter m to the size of the blind spot is revealed by the behavior of the ratio $(m-1)/(m+1)$ viewed as a function of m. Since $(m-1)/(m+1)$ increases as the value of m increases, increasing the value of m expands the blind spot. For a fixed value of $m > 1$, $B_{\text{lin}} = 0$ only if $b = 0$. For $b \neq 0$, B_{lin} decreases and approaches 0 as a increases and approaches infinity. If $a = 1$ and $b = \gamma^2$, B_{lin} approaches 0 as γ approaches 0. If $a = \delta$ and $b = 1$, B_{lin} decreases and approaches 0 as δ increases and approaches infinity.

The blind spot $\mathcal{R}_{B_{\text{exp}}}$ corresponding to the exponential generator function $g_0(x) = \exp(\beta x)$ is determined by:

$$B_{\text{exp}} = \frac{m-1}{2\beta}. \tag{39}$$

For a fixed value of β, the blind spot depends exclusively on the parameter m. Once again, the blind spot corresponding to the exponential generator function expands as the value of m increases. For a fixed value of $m > 1$, B_{exp} decreases and approaches 0 as β increases and approaches infinity. For $g_0(x) = \exp(\beta x)$, $g(x) = (g_0(x))^{\frac{1}{1-m}} = \exp(-x/\sigma^2)$ with $\sigma^2 = (m-1)/\beta$. As a result, the blind spot corresponding to the exponential generator function approaches 0 only if the width of the Gaussian radial basis function $\phi(x) = g(x^2) = \exp(-x^2/\sigma^2)$ approaches 0. Such a range of values of σ would make it difficult for Gaussian radial basis functions to behave as receptive fields that can cover the entire input space.

It is clear from (38) and (39) that the blind spot corresponding to the exponential generator function is much more sensitive to changes of m compared with that corresponding to the linear generator function. This can be quantified by computing for both generator functions the relative sensitivity of $B = B(m)$ in terms of m, defined as:

$$S_B^m = \frac{m}{B}\frac{\partial B}{\partial m}. \tag{40}$$

For the linear generator function $g_0(x) = ax + b$, $\partial B_{\text{lin}}/\partial m = (2/(m+1)^2)(b/a)$ and

$$S_{B_{\text{lin}}}^m = \frac{2m}{m^2 - 1}. \tag{41}$$

For the exponential generator function $g_0(x) = \exp(\beta x)$, $\partial B_{\text{exp}}/\partial m = 1/(2\beta)$ and

$$S_{B_{\text{exp}}}^m = \frac{m}{m-1}. \tag{42}$$

Combining (41) and (42) gives:

$$S_{B_{\text{exp}}}^m = \frac{m+1}{2} S_{B_{\text{lin}}}^m. \tag{43}$$

Since $m > 1$, $S_{B_{\text{exp}}}^m > S_{B_{\text{lin}}}^m$. As an example, for $m = 3$ the sensitivity with respect to m of the blind spot corresponding to the exponential generator function is twice that corresponding to the linear generator function.

5.2   Criteria for Selecting Generator Functions

The response of the radial basis function located at the prototype $\mathbf{v}_j$ to training vectors depends on their Euclidean distance from $\mathbf{v}_j$ and the shape of the generator function used. If the generator function does not satisfy the fifth axiomatic requirement, the response of the radial basis function located at each prototype exhibits the desired behavior only if the training vectors are located outside its blind spot. This implies that the training of a RBF model by a learning procedure based on gradient descent depends mainly on the sensitivity of the radial basis functions to training vectors outside their blind spots. This indicates that the criteria used for selecting generator functions should involve both the shapes of the radial basis functions relative to their blind spots and the sensitivity of the radial basis functions to input vectors outside their blind spots. The sensitivity of the response $h_{j,k}$ of the jth radial basis function to any input $\mathbf{x}_k$ can be measured by the norm of the gradient $\nabla_{\mathbf{x}_k} h_{j,k}$. Thus, the shape and sensitivity of the radial basis function located at the prototype $\mathbf{v}_j$ are mainly affected by:

1. the value $h_j^* = g(B)$ of the response $h_{j,k} = g(\|\mathbf{x}_k - \mathbf{v}_j\|^2)$ of the jth radial basis function at $\|\mathbf{x}_k - \mathbf{v}_j\|^2 = B$ and the rate at which $h_{j,k} = g(\|\mathbf{x}_k - \mathbf{v}_j\|^2)$ decreases as $\|\mathbf{x}_k - \mathbf{v}_j\|^2$ increases above B and approaches infinity, and

2. the maximum value attained by the norm of the gradient $\nabla_{\mathbf{x}_k} h_{j,k}$ at $\|\mathbf{x}_k - \mathbf{v}_j\|^2 = B$ and the rate at which $\|\nabla_{\mathbf{x}_k} h_{j,k}\|^2$ decreases as $\|\mathbf{x}_k - \mathbf{v}_j\|^2$ increases above B and approaches infinity.

The criteria that may be used for selecting radial basis functions can be established by considering the following extreme situation. Suppose the response $h_{j,k} = g(\|\mathbf{x}_k - \mathbf{v}_j\|^2)$ diminishes very quickly and the receptive field located at the prototype $\mathbf{v}_j$ does not extend far beyond the blind spot. This can have a negative impact on the function approximation ability of the corresponding RBF model since the region outside the blind spot contains the input vectors that affect the implementation of the input-output mapping as indicated by the sensitivity measure $\|\nabla_{\mathbf{x}_k} h_{j,k}\|^2$. Thus, a generator function must be selected in such a way that:

1. the response $h_{j,k}$ and the sensitivity measure $\|\nabla_{\mathbf{x}_k} h_{j,k}\|^2$ take substantial values outside the blind spot before they approach 0, and

2. the response $h_{j,k}$ is sizable outside the blind sport even after the values of $\|\nabla_{\mathbf{x}_k} h_{j,k}\|^2$ become negligible.

The rate at which the response $h_{j,k} = g(\|\mathbf{x}_k - \mathbf{v}_j\|^2)$ decreases relates to the "tails" of the functions $g(\cdot)$ that correspond to different generator functions. The use of a short-tailed function $g(\cdot)$ shrinks the receptive fields of the RBF model while the use of a long-tailed function $g(\cdot)$ increases the overlapping between the receptive fields located at different prototypes. If $g(x) = (g_0(x))^{\frac{1}{1-m}}$ and $m > 1$, the tail of $g(x)$ is determined by how fast the corresponding generator function $g_0(x)$ changes as a function of x. As x increases, the exponential generator function $g_0(x) = \exp(\beta x)$ increases faster than the linear generator function $g_0(x) = a x + b$. As a result, the response $g(x) = (g_0(x))^{\frac{1}{1-m}}$ diminishes quickly if $g_0(\cdot)$ is exponential and slower if $g_0(\cdot)$ is linear.

The behavior of the sensitivity measure $\|\nabla_{\mathbf{x}_k} h_{j,k}\|^2$ also depends on the properties of the function $g(\cdot)$. For $h_{j,k} = g(\|\mathbf{x}_k - \mathbf{v}_j\|^2)$, $\nabla_{\mathbf{x}_k} h_{j,k}$ can be obtained from (35) as:

$$\nabla_{\mathbf{x}_k} h_{j,k} = -\alpha_{j,k} \left(\mathbf{x}_k - \mathbf{v}_j \right), \tag{44}$$

where

$$\alpha_{j,k} = -2\,g'(\|\mathbf{x}_k - \mathbf{v}_j\|^2). \tag{45}$$

From (44),

$$\|\nabla_{\mathbf{x}_k} h_{j,k}\|^2 = \|\mathbf{x}_k - \mathbf{v}_j\|^2 \, \alpha_{j,k}^2. \tag{46}$$

The selection of a specific function $g(\cdot)$ influences the sensitivity measure $\|\nabla_{\mathbf{x}_k} h_{j,k}\|^2$ through $\alpha_{j,k} = -2\,g'(\|\mathbf{x}_k - \mathbf{v}_j\|^2)$. If $g(x) = (g_0(x))^{\frac{1}{1-m}}$, then:

$$\begin{aligned}
g'(x) &= \frac{1}{1-m} \, (g_0(x))^{\frac{m}{1-m}} \, g_0'(x) \\
&= \frac{1}{1-m} \, (g(x))^m \, g_0'(x).
\end{aligned} \tag{47}$$

Since $h_{j,k} = g\left(\|\mathbf{x}_k - \mathbf{v}_j\|^2\right)$, $\alpha_{j,k}$ is given by:

$$\alpha_{j,k} = \frac{2}{m-1} \, (h_{j,k})^m \, g_0'(\|\mathbf{x}_k - \mathbf{v}_j\|^2). \tag{48}$$

5.3   Evaluation of Linear and Exponential Generator Functions

The criteria presented above are used here for evaluating linear and exponential generator functions.

5.3.1   Linear Generator Functions

If $g(x) = (g_0(x))^{\frac{1}{1-m}}$, with $g_0(x) = a\,x + b$ and $m > 1$, the response $h_{j,k} = g(\|\mathbf{x}_k - \mathbf{v}_j\|^2)$ of the jth radial basis function to $\mathbf{x}_k$ is:

$$h_{j,k} = \left(\frac{1}{a\,\|\mathbf{x}_k - \mathbf{v}_j\|^2 + b} \right)^{\frac{1}{m-1}}. \tag{49}$$

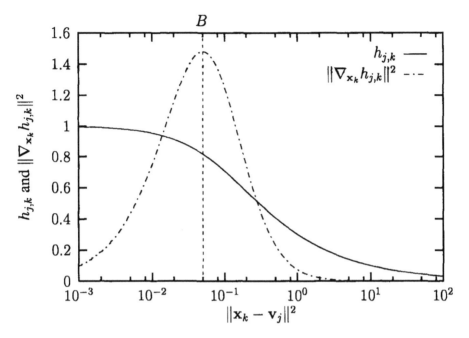

Figure 2. The response $h_{j,k} = g(\|\mathbf{x}_k - \mathbf{v}_j\|^2)$ of the jth radial basis function and the norm of the gradient $\|\nabla_{\mathbf{x}_k} h_{j,k}\|^2$ plotted as functions of $\|\mathbf{x}_k - \mathbf{v}_j\|^2$ for $g(x) = (g_0(x))^{\frac{1}{1-m}}$, with $g_0(x) = 1 + \delta x$, $m = 3$, and $\delta = 10$.

For this generator function, $g_0'(x) = a$ and (48) gives:

$$\alpha_{j,k} = \frac{2a}{m-1}(h_{j,k})^m$$

$$= \frac{2a}{m-1}\left(\frac{1}{a\|\mathbf{x}_k - \mathbf{v}_j\|^2 + b}\right)^{\frac{m}{m-1}}. \tag{50}$$

Thus, $\|\nabla_{\mathbf{x}_k} h_{j,k}\|^2$ can be obtained from (46) as:

$$\|\nabla_{\mathbf{x}_k} h_{j,k}\|^2 = \left(\frac{2a}{m-1}\right)^2 \|\mathbf{x}_k - \mathbf{v}_j\|^2 \left(\frac{1}{a\|\mathbf{x}_k - \mathbf{v}_j\|^2 + b}\right)^{\frac{2m}{m-1}}. \tag{51}$$

Figures 2 and 3 show the response $h_{j,k} = g(\|\mathbf{x}_k - \mathbf{v}_j\|^2)$ of the jth radial basis function to the input vector $\mathbf{x}_k$ and the sensitivity measure $\|\nabla_{\mathbf{x}_k} h_{j,k}\|^2$ plotted as functions of $\|\mathbf{x}_k - \mathbf{v}_j\|^2$ for $g(x) = (g_0(x))^{\frac{1}{1-m}}$, with $g_0(x) = 1 + \delta x$, $m = 3$, for $\delta = 10$ and $\delta = 100$, respectively. In accordance with the analysis, $\|\nabla_{\mathbf{x}_k} h_{j,k}\|^2$ increases monotonically as

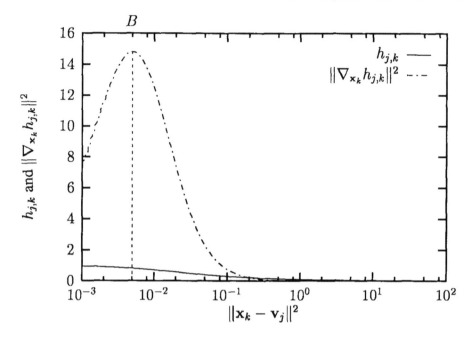

Figure 3. The response $h_{j,k} = g(\|\mathbf{x}_k - \mathbf{v}_j\|^2)$ of the jth radial basis function and the norm of the gradient $\|\nabla_{\mathbf{x}_k} h_{j,k}\|^2$ plotted as functions of $\|\mathbf{x}_k - \mathbf{v}_j\|^2$ for $g(x) = (g_0(x))^{\frac{1}{1-m}}$, with $g_0(x) = 1 + \delta x$, $m = 3$, and $\delta = 100$.

$\|\mathbf{x}_k - \mathbf{v}_j\|^2$ increases from 0 to $B = 1/(2\,\delta)$ and decreases monotonically as $\|\mathbf{x}_k - \mathbf{v}_j\|^2$ increases above B and approaches infinity. Figures 2 and 3 indicate that, regardless of the value of δ, the response $h_{j,k}$ of the radial basis function located at the prototype $\mathbf{v}_j$ is sizable outside the blind spot even after the values of $\|\nabla_{\mathbf{x}_k} h_{j,k}\|^2$ become negligible. Thus, the radial basis function located at the prototype $\mathbf{v}_j$ is activated by all input vectors that correspond to substantial values of $\|\nabla_{\mathbf{x}_k} h_{j,k}\|^2$.

5.3.2   Exponential Generator Functions

If $g(x) = (g_0(x))^{\frac{1}{1-m}}$, with $g_0(x) = \exp(\beta x)$ and $m > 1$, the response $h_{j,k} = g(\|\mathbf{x}_k - \mathbf{v}_j\|^2)$ of the jth radial basis function to $\mathbf{x}_k$ is:

$$h_{j,k} = \exp\left(-\frac{\|\mathbf{x}_k - \mathbf{v}_j\|^2}{\sigma^2}\right), \tag{52}$$

where $\sigma^2 = (m - 1)/\beta$. For this generator function, $g_0'(x) = \beta \exp(\beta x) = \beta\, g_0(x)$. In this case, $g_0'(\|\mathbf{x}_k - \mathbf{v}_j\|^2) = \beta\,(h_{j,k})^{1-m}$ and

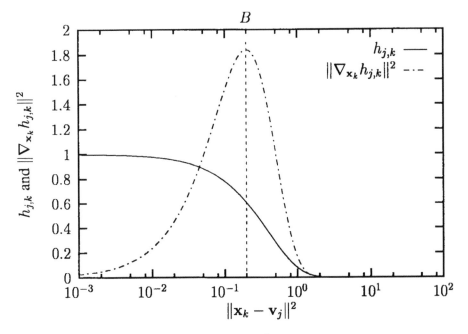

Figure 4. The response $h_{j,k} = g(\|\mathbf{x}_k - \mathbf{v}_j\|^2)$ of the jth radial basis function and the norm of the gradient $\|\nabla_{\mathbf{x}_k} h_{j,k}\|^2$ plotted as functions of $\|\mathbf{x}_k - \mathbf{v}_j\|^2$ for $g(x) = (g_0(x))^{\frac{1}{1-m}}$, with $g_0(x) = \exp(\beta x)$, $m = 3$, and $\beta = 5$.

(48) gives:

$$
\begin{aligned}
\alpha_{j,k} &= \frac{2\beta}{m-1} h_{j,k} \\
&= \frac{2}{\sigma^2} \exp\left(-\frac{\|\mathbf{x}_k - \mathbf{v}_j\|^2}{\sigma^2}\right).
\end{aligned} \tag{53}
$$

Thus, $\|\nabla_{\mathbf{x}_k} h_{j,k}\|^2$ can be obtained from (46) as:

$$
\|\nabla_{\mathbf{x}_k} h_{j,k}\|^2 = \left(\frac{2}{\sigma^2}\right)^2 \|\mathbf{x}_k - \mathbf{v}_j\|^2 \exp\left(-2\frac{\|\mathbf{x}_k - \mathbf{v}_j\|^2}{\sigma^2}\right). \tag{54}
$$

Figures 4 and 5 show the response $h_{j,k} = g(\|\mathbf{x}_k - \mathbf{v}_j\|^2)$ of the jth radial basis function to the input vector $\mathbf{x}_k$ and the sensitivity measure $\|\nabla_{\mathbf{x}_k} h_{j,k}\|^2$ plotted as functions of $\|\mathbf{x}_k - \mathbf{v}_j\|^2$ for $g(x) = (g_0(x))^{\frac{1}{1-m}}$, with $g_0(x) = \exp(\beta x)$, $m = 3$, for $\beta = 5$ and $\beta = 10$, respectively. Once again, $\|\nabla_{\mathbf{x}_k} h_{j,k}\|^2$ increases monotonically as $\|\mathbf{x}_k - \mathbf{v}_j\|^2$ increases from 0 to $B = 1/\beta$ and decreases monotonically as $\|\mathbf{x}_k - \mathbf{v}_j\|^2$ increases

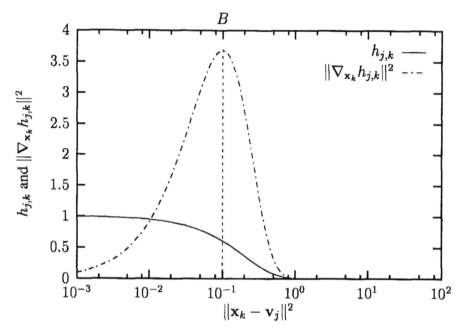

Figure 5. The response $h_{j,k} = g(\|\mathbf{x}_k - \mathbf{v}_j\|^2)$ of the jth radial basis function and the norm of the gradient $\|\nabla_{\mathbf{x}_k} h_{j,k}\|^2$ plotted as functions of $\|\mathbf{x}_k - \mathbf{v}_j\|^2$ for $g(x) = (g_0(x))^{\frac{1}{1-m}}$, with $g_0(x) = \exp(\beta x)$, $m = 3$, and $\beta = 10$.

above B and approaches infinity. Nevertheless, there are some significant differences between the response $h_{j,k}$ and the sensitivity measure
$\|\nabla_{\mathbf{x}_k} h_{j,k}\|^2$ corresponding to linear and exponential generator functions.
If $g_0(x) = \exp(\beta x)$, then the response $h_{j,k}$ is substantial for the input
vectors inside the blind spot but diminishes very quickly for values of
$\|\mathbf{x}_k - \mathbf{v}_j\|^2$ above B. In fact, the values of $h_{j,k}$ become negligible even
before $\|\nabla_{\mathbf{x}_k} h_{j,k}\|^2$ approaches asymptotically zero values. This is in di-
rect contrast with the behavior of the same quantities corresponding to
linear generator functions, which are shown in Figures 2 and 3.

6   Learning Algorithms Based on Gradient Descent

Reformulated RBF neural networks can be trained to map $\mathbf{x}_k \in \mathbf{R}^{n_i}$ into
$\mathbf{y}_k = [y_{1,k}\, y_{2,k}\, \cdots\, y_{n_o,k}]^T \in \mathbf{R}^{n_o}$, where the vector pairs $(\mathbf{x}_k, \mathbf{y}_k)$, $1 \leq$
$k \leq M$, form the training set. If $\mathbf{x}_k \in \mathbf{R}^{n_i}$ is the input to a reformulated
RBF network, its response is $\hat{\mathbf{y}}_k = [\hat{y}_{1,k}\, \hat{y}_{2,k}\, \cdots\, \hat{y}_{n_o,k}]^T$, where $\hat{y}_{i,k}$ is the

actual response of the ith output unit to $\mathbf{x}_k$ given by:

$$
\begin{aligned}
\hat{y}_{i,k} &= f(\bar{y}_{i,k}) \\
&= f\left(\mathbf{w}_i^T \mathbf{h}_k\right) \\
&= f\left(\sum_{j=0}^{c} w_{ij} h_{j,k}\right),
\end{aligned}
\tag{55}
$$

with $h_{0,k} = 1$, $1 \le k \le M$, $h_{j,k} = g\left(\|\mathbf{x}_k - \mathbf{v}_j\|^2\right)$, $1 \le j \le c$, $\mathbf{h}_k = [h_{0,k}\, h_{1,k}\, \ldots\, h_{c,k}]^T$, and $\mathbf{w}_i = [w_{i,0}\, w_{i,1}\, \ldots\, w_{i,c}]^T$. Training is typically based on the minimization of the error between the actual outputs of the network $\hat{\mathbf{y}}_k$, $1 \le k \le M$, and the desired responses $\mathbf{y}_k$, $1 \le k \le M$.

6.1   Batch Learning Algorithms

A reformulated RBF neural network can be trained by minimizing the error:

$$
E = \frac{1}{2} \sum_{k=1}^{M} \sum_{i=1}^{n_o} (y_{i,k} - \hat{y}_{i,k})^2.
\tag{56}
$$

Minimization of (56) using gradient descent implies that all training examples are presented to the RBF network simultaneously. Such training strategy leads to *batch* learning algorithms. The update equation for the weight vectors of the upper associative network can be obtained using gradient descent as [21]:

$$
\begin{aligned}
\Delta \mathbf{w}_p &= -\eta \, \nabla_{\mathbf{w}_p} E \\
&= \eta \sum_{k=1}^{M} \varepsilon_{p,k}^o \, \mathbf{h}_k,
\end{aligned}
\tag{57}
$$

where η is the learning rate and $\varepsilon_{p,k}^o$ is the *output error*, given as:

$$
\varepsilon_{p,k}^o = f'(\bar{y}_{p,k})\, (y_{p,k} - \hat{y}_{p,k}).
\tag{58}
$$

Similarly, the update equation for the prototypes can be obtained using gradient descent as [21]:

$$
\begin{aligned}
\Delta \mathbf{v}_q &= -\eta \, \nabla_{\mathbf{v}_q} E \\
&= \eta \sum_{k=1}^{M} \varepsilon_{q,k}^h \, (\mathbf{x}_k - \mathbf{v}_q),
\end{aligned}
\tag{59}
$$

where η is the learning rate and $\varepsilon_{q,k}^h$ is the *hidden error*, defined as:

$$\varepsilon_{q,k}^h = \alpha_{q,k} \sum_{i=1}^{n_o} \varepsilon_{i,k}^o \, w_{iq}, \tag{60}$$

with $\alpha_{q,k} = -2\,g'\left(\|\mathbf{x}_k - \mathbf{v}_q\|^2\right)$. The selection of a specific function $g(\cdot)$ influences the update of the prototypes through $\alpha_{q,k} = -2\,g'(\|\mathbf{x}_k - \mathbf{v}_q\|^2)$, which is involved in the calculation of the corresponding hidden error $\varepsilon_{q,k}^h$. Since $h_{q,k} = g\left(\|\mathbf{x}_k - \mathbf{v}_q\|^2\right)$ and $g(x) = (g_0(x))^{\frac{1}{1-m}}$, $\alpha_{q,k}$ is given by (48) and the hidden error (60) becomes:

$$\varepsilon_{q,k}^h = \frac{2}{m-1}\,(h_{q,k})^m\,g_0'(\|\mathbf{x}_k - \mathbf{v}_q\|^2) \sum_{i=1}^{n_o} \varepsilon_{i,k}^o \, w_{iq}. \tag{61}$$

A RBF neural network can be trained according to the algorithm presented above in a sequence of *adaptation cycles*, where an adaptation cycle involves the update of all adjustable parameters of the network. An adaptation cycle begins by replacing the current estimate of each weight vector $\mathbf{w}_p, 1 \le p \le n_o$, by its updated version:

$$\mathbf{w}_p + \Delta\mathbf{w}_p = \mathbf{w}_p + \eta \sum_{k=1}^{M} \varepsilon_{p,k}^o \, \mathbf{h}_k. \tag{62}$$

Given the learning rate η and the responses $\mathbf{h}_k$ of the radial basis functions, these weight vectors are updated according to the output errors $\varepsilon_{p,k}^o, 1 \le p \le n_o$. Following the update of these weight vectors, the current estimate of each prototype $\mathbf{v}_q, 1 \le q \le c$, is replaced by:

$$\mathbf{v}_q + \Delta\mathbf{v}_q = \mathbf{v}_q + \eta \sum_{k=1}^{M} \varepsilon_{q,k}^h \, (\mathbf{x}_k - \mathbf{v}_q). \tag{63}$$

For a given value of the learning rate η, the update of $\mathbf{v}_q$ depends on the hidden errors $\varepsilon_{q,k}^h, 1 \le k \le M$. The hidden error $\varepsilon_{q,k}^h$ is influenced by the output errors $\varepsilon_{i,k}^o, 1 \le i \le n_o$, and the weights $w_{iq}, 1 \le i \le n_o$, through the term $\sum_{i=1}^{n_o} \varepsilon_{i,k}^o \, w_{iq}$. Thus, the RBF neural network is trained according to this scheme by propagating back the output error.

This algorithm can be summarized as follows:

1. Select η and ϵ; initialize $\{w_{ij}\}$ with zero values; initialize the prototypes $\mathbf{v}_j, 1 \le j \le c$; set $h_{0,k} = 1, \forall k$.

2. Compute the initial response:

- $h_{j,k} = (g_0 (\|\mathbf{x}_k - \mathbf{v}_j\|^2))^{\frac{1}{1-m}}, \forall j, k.$
- $\mathbf{h}_k = [h_{0,k}\, h_{1,k}\, \ldots\, h_{c,k}]^T, \forall k.$
- $\hat{y}_{i,k} = f(\mathbf{w}_i^T \mathbf{h}_k), \forall i, k.$

3. Compute $E = \frac{1}{2}\sum_{k=1}^{M}\sum_{i=1}^{n_o}(y_{i,k} - \hat{y}_{i,k})^2.$

4. Set $E_{\text{old}} = E.$

5. Update the adjustable parameters:

- $\varepsilon_{i,k}^o = f'(\bar{y}_{i,k})(y_{i,k} - \hat{y}_{i,k}), \forall i, k.$
- $\mathbf{w}_i \leftarrow \mathbf{w}_i + \eta \sum_{k=1}^{M} \varepsilon_{i,k}^o \mathbf{h}_k, \forall i.$
- $\varepsilon_{j,k}^h = \frac{2}{m-1} g_0' (\|\mathbf{x}_k - \mathbf{v}_j\|^2)(h_{j,k})^m \sum_{i=1}^{n_o} \varepsilon_{i,k}^o w_{ij}, \forall j, k.$
- $\mathbf{v}_j \leftarrow \mathbf{v}_j + \eta \sum_{k=1}^{M} \varepsilon_{j,k}^h (\mathbf{x}_k - \mathbf{v}_j), \forall j.$

6. Compute the current response:

- $h_{j,k} = (g_0 (\|\mathbf{x}_k - \mathbf{v}_j\|^2))^{\frac{1}{1-m}}, \forall j, k.$
- $\mathbf{h}_k = [h_{0,k}\, h_{1,k}\, \ldots\, h_{c,k}]^T, \forall k.$
- $\hat{y}_{i,k} = f(\mathbf{w}_i^T \mathbf{h}_k), \forall i, k.$

7. Compute $E = \frac{1}{2}\sum_{k=1}^{M}\sum_{i=1}^{n_o}(y_{i,k} - \hat{y}_{i,k})^2.$

8. If: $(E_{\text{old}} - E)/E_{\text{old}} > \epsilon$; then: go to 4.

6.2 Sequential Learning Algorithms

Reformulated RBF neural networks can also be trained "on-line" by *sequential* learning algorithms. Such algorithms can be developed by using gradient descent to minimize the errors:

$$E_k = \frac{1}{2}\sum_{i=1}^{n_o}(y_{i,k} - \hat{y}_{i,k})^2, \qquad (64)$$

for $k = 1, 2, \ldots, M$. The update equation for the weight vectors of the upper associative network can be obtained using gradient descent as [21]:

$$
\begin{aligned}
\Delta \mathbf{w}_{p,k} &= \mathbf{w}_{p,k} - \mathbf{w}_{p,k-1} \\
&= -\eta \nabla_{\mathbf{w}_p} E_k \\
&= \eta \varepsilon^o_{p,k} \mathbf{h}_k,
\end{aligned}
\tag{65}
$$

where $\mathbf{w}_{p,k-1}$ and $\mathbf{w}_{p,k}$ are the estimates of the weight vector $\mathbf{w}_p$ before and after the presentation of the training example $(\mathbf{x}_k, \mathbf{y}_k)$, η is the learning rate, and $\varepsilon^o_{p,k}$ is the output error defined in (58). Similarly, the update equation for the prototypes can be obtained using gradient descent as [21]:

$$
\begin{aligned}
\Delta \mathbf{v}_{q,k} &= \mathbf{v}_{q,k} - \mathbf{v}_{q,k-1} \\
&= -\eta \nabla_{\mathbf{v}_q} E_k \\
&= \eta \varepsilon^h_{q,k} (\mathbf{x}_k - \mathbf{v}_q),
\end{aligned}
\tag{66}
$$

where $\mathbf{v}_{q,k-1}$ and $\mathbf{v}_{q,k}$ are the estimates of the prototype $\mathbf{v}_q$ before and after the presentation of the training example $(\mathbf{x}_k, \mathbf{y}_k)$, η is the learning rate, and $\varepsilon^h_{q,k}$ is the hidden error defined in (61).

When an adaptation cycle begins, the current estimates of the weight vectors $\mathbf{w}_p$ and the prototypes $\mathbf{v}_q$ are stored in $\mathbf{w}_{p,0}$ and $\mathbf{v}_{q,0}$, respectively. After an example $(\mathbf{x}_k, \mathbf{y}_k)$, $1 \leq k \leq M$, is presented to the network, each weight vector $\mathbf{w}_p$, $1 \leq p \leq n_o$, is updated as:

$$
\mathbf{w}_{p,k} = \mathbf{w}_{p,k-1} + \Delta \mathbf{w}_{p,k} = \mathbf{w}_{p,k-1} + \eta \varepsilon^o_{p,k} \mathbf{h}_k.
\tag{67}
$$

Following the update of all the weight vectors $\mathbf{w}_p$, $1 \leq p \leq n_o$, each prototype $\mathbf{v}_q$, $1 \leq q \leq c$, is updated as:

$$
\mathbf{v}_{q,k} = \mathbf{v}_{q,k-1} + \Delta \mathbf{v}_{q,k} = \mathbf{v}_{q,k-1} + \eta \varepsilon^h_{q,k} (\mathbf{x}_k - \mathbf{v}_{q,k-1}).
\tag{68}
$$

An adaptation cycle is completed in this case after the sequential presentation to the network of all the examples included in the training set. Once again, the RBF neural network is trained according to this scheme by propagating back the output error.

This algorithm can be summarized as follows:

1. Select η and ϵ; initialize $\{w_{ij}\}$ with zero values; initialize the prototypes $\mathbf{v}_j$, $1 \leq j \leq c$; set $h_{0,k} = 1, \forall k$.

2. Compute the initial response:

- $h_{j,k} = (g_0 (\|\mathbf{x}_k - \mathbf{v}_j\|^2))^{\frac{1}{1-m}}$, $\forall j, k$.
- $\mathbf{h}_k = [h_{0,k}\, h_{1,k}\, \ldots\, h_{c,k}]^T$, $\forall k$.
- $\hat{y}_{i,k} = f(\mathbf{w}_i^T \mathbf{h}_k)$, $\forall i, k$.

3. Compute $E = \frac{1}{2} \sum_{k=1}^{M} \sum_{i=1}^{n_o} (y_{i,k} - \hat{y}_{i,k})^2$.

4. Set $E_{\text{old}} = E$.

5. Update the adjustable parameters for all $k = 1, 2, \ldots, M$:

- $\varepsilon_{i,k}^o = f'(\bar{y}_{i,k})(y_{i,k} - \hat{y}_{i,k})$, $\forall i$.
- $\mathbf{w}_i \leftarrow \mathbf{w}_i + \eta\, \varepsilon_{i,k}^o\, \mathbf{h}_k$, $\forall i$.
- $\varepsilon_{j,k}^h = \frac{2}{m-1} g_0' (\|\mathbf{x}_k - \mathbf{v}_j\|^2) (h_{j,k})^m \sum_{i=1}^{n_o} \varepsilon_{i,k}^o\, w_{ij}$, $\forall j$.
- $\mathbf{v}_j \leftarrow \mathbf{v}_j + \eta\, \varepsilon_{j,k}^h (\mathbf{x}_k - \mathbf{v}_j)$, $\forall j$.

6. Compute the current response:

- $h_{j,k} = (g_0 (\|\mathbf{x}_k - \mathbf{v}_j\|^2))^{\frac{1}{1-m}}$, $\forall j, k$.
- $\mathbf{h}_k = [h_{0,k}\, h_{1,k}\, \ldots\, h_{c,k}]^T$, $\forall k$.
- $\hat{y}_{i,k} = f(\mathbf{w}_i^T \mathbf{h}_k)$, $\forall i, k$.

7. Compute $E = \frac{1}{2} \sum_{k=1}^{M} \sum_{i=1}^{n_o} (y_{i,k} - \hat{y}_{i,k})^2$.

8. If: $(E_{\text{old}} - E)/E_{\text{old}} > \epsilon$; then: go to 4.

6.3 Initialization of Supervised Learning

The training of reformulated RBF neural networks using gradient descent can be initialized by randomly generating the set of prototypes that determine the locations of the radial basis function centers in the input space. Such an approach relies on the supervised learning algorithm to determine appropriate locations for the radial basis function centers by updating the prototypes during learning. Nevertheless, the training of reformulated RBF neural networks by gradient descent algorithms can be facilitated by initializing the supervised learning process using a set of prototypes specifically determined to represent the input vectors included in the training set. This can be accomplished by computing the initial set of prototypes using unsupervised clustering or learning vector quantization (LVQ) algorithms.

According to the learning scheme often used for training conventional RBF neural networks [34], the locations of the radial basis function centers are determined from the input vectors included in the training set using the c-means (or k-means) algorithm. The c-means algorithm begins from an initial set of c prototypes, which implies the partition of the input vectors into c clusters. Each cluster is represented by a prototype, which is evaluated at subsequent iterations as the centroid of the input vectors belonging to that cluster. Each input vector is assigned to the cluster whose prototype is its closest neighbor. In mathematical terms, the indicator function $u_{ij} = u_j(\mathbf{x}_i)$ that assigns the input vector $\mathbf{x}_i$ to the jth cluster is computed as [9]:

$$u_{ij} = \begin{cases} 1, & \text{if } \|\mathbf{x}_i - \mathbf{v}_j\|^2 < \|\mathbf{x}_i - \mathbf{v}_\ell\|^2, \forall \ell \neq j, \\ 0, & \text{otherwise.} \end{cases} \tag{69}$$

For a given set of indicator functions, the new set of prototypes is calculated as [9]:

$$\mathbf{v}_j = \frac{\sum_{i=1}^{M} u_{ij}\,\mathbf{x}_i}{\sum_{i=1}^{M} u_{ij}}, \quad 1 \leq j \leq c. \tag{70}$$

The c-means algorithm partitions the input vectors into clusters represented by a set of prototypes based on *hard* or *crisp* decisions. In other words, each input vector is assigned to the cluster represented by its closest prototype. Since this strategy fails to quantify the uncertainty typically associated with partitioning a set of input vectors, the performance of the c-means algorithm depends rather strongly on its initialization [8], [26]. When this algorithm is initialized randomly, it often converges to shallow local minima and produces empty clusters.

Most of the disadvantages of the c-means algorithm can be overcome by employing a prototype splitting procedure to produce the initial set of prototypes. Such a procedure is employed by a variation of the c-means algorithm often referred to in the literature as the LBG (Linde-Buzo-Gray) algorithm [31], which is often used for codebook design in image and video compression approaches based on vector quantization.

The LBG algorithm employs an initialization scheme to compensate for the dependence of the c-means algorithm on its initialization [8]. More specifically, this algorithm generates the desired number of clusters by successively splitting the prototypes and subsequently employing the c-means algorithm. The algorithm begins with a single prototype that is calculated as the centroid of the available input vectors. This prototype is split into two vectors, which provide the initial estimate for the c-means algorithm that is used with $c = 2$. Each of the resulting vectors is then split into two vectors and the above procedure is repeated until the desired number of prototypes is obtained. Splitting is performed by adding the perturbation vectors $\pm e_i$ to each vector v_i producing two vectors: $v_i + e_i$ and $v_i - e_i$. The perturbation vector e_i can be calculated from the variance between the input vectors and the prototypes [8].

7 Generator Functions and Gradient Descent Learning

The effect of the generator function on gradient descent learning algorithms developed for reformulated RBF neural networks essentially relates to the criteria established in Section 5 for selecting generator functions. These criteria were established on the basis of the response $h_{j,k}$ of the jth radial basis function to an input vector x_k and the norm of the gradient $\nabla_{x_k} h_{j,k}$, that can be used to measure the sensitivity of the radial basis function response $h_{j,k}$ to an input vector x_k. Since $\nabla_{x_k} h_{j,k} = -\nabla_{v_j} h_{j,k}$, (46) gives

$$\|\nabla_{v_j} h_{j,k}\|^2 = \|x_k - v_j\|^2 \, \alpha_{j,k}^2. \tag{71}$$

According to (71), the quantity $\|x_k - v_j\|^2 \, \alpha_{j,k}^2$ can also be used to measure the sensitivity of the response of the jth radial basis function to changes in the prototype v_j that represents its location in the input space.

The gradient descent learning algorithms presented in Section 6 attempt to train a RBF neural network to implement a desired input-output mapping by producing incremental changes of its adjustable parameters, i.e., the output weights and the prototypes. If the responses of the radial basis functions are not substantially affected by incremental changes of the prototypes, then the learning process reduces to incremental changes of the output weights and eventually the algorithm trains a single-layered neural network. Given the limitations of single-layered neural networks [28], such updates alone are unlikely to implement non-trivial input-output mappings. Thus, the ability of the network to implement a desired input-output mapping depends to a large extent on the sensitivity of the responses of the radial basis functions to incremental changes of their corresponding prototypes. This discussion indicates that the sensitivity measure $\|\nabla_{\mathbf{v}_j} h_{j,k}\|^2$ is relevant to gradient descent learning algorithms developed for reformulated RBF neural networks. Moreover, the form of this sensitivity measure in (71) underlines the significant role of the generator function, whose selection affects $\|\nabla_{\mathbf{v}_j} h_{j,k}\|^2$ as indicated by the definition of $\alpha_{j,k}$ in (48). The effect of the generator function on gradient descent learning is revealed by comparing the response $h_{j,k}$ and the sensitivity measure $\|\nabla_{\mathbf{v}_j} h_{j,k}\|^2 = \|\nabla_{\mathbf{x}_k} h_{j,k}\|^2$ corresponding to the linear and exponential generator functions.

According to Figures 2 and 3, the response $h_{j,k}$ of the jth radial basis function to the input $\mathbf{x}_k$ diminishes very slowly outside the blind spot, i.e., as $\|\mathbf{x}_k - \mathbf{v}_j\|^2$ increases above B. This implies that the training vector $\mathbf{x}_k$ has a non-negligible effect on the response $h_{j,k}$ of the radial basis function located at this prototype. The behavior of the sensitivity measure $\|\nabla_{\mathbf{v}_j} h_{j,k}\|^2$ outside the blind spot indicates that the update of the prototype $\mathbf{v}_j$ produces significant variations in the input of the upper associative network, which is trained to implement the desired input-output mapping by updating the output weights. Figures 2 and 3 also reveal the trade-off involved in the selection of the free parameter δ in practice. As the value of δ increases, $\|\nabla_{\mathbf{v}_j} h_{j,k}\|^2$ attains significantly higher values. This implies that the jth radial basis function is more sensitive to updates of the prototype $\mathbf{v}_j$ due to input vectors outside its blind spot. The blind spot shrinks as the value of δ increases but $\|\nabla_{\mathbf{v}_j} h_{j,k}\|^2$ approaches 0 quickly outside the blind spot, i.e., as the value of $\|\mathbf{x}_k - \mathbf{v}_j\|^2$ increases above B. This implies that the receptive fields located at the prototypes shrink, which can have a negative impact on the gradient descent learning. Increasing the value of δ can also affect the number of radial basis functions required for the implementation of the desired input-output mapping. This is due to the fact that more radial basis functions are required to cover the input space. The receptive fields located at the prototypes can be expanded by decreasing the value of δ. However, $\|\nabla_{\mathbf{v}_j} h_{j,k}\|^2$

becomes flat as the value of δ decreases. This implies that very small values of δ can decrease the sensitivity of the radial basis functions to the input vectors included in their receptive fields.

According to Figures 4 and 5, the response of the jth radial basis function to the input $\mathbf{x}_k$ diminishes very quickly outside the blind spot, i.e., as $\|\mathbf{x}_k - \mathbf{v}_j\|^2$ increases above B. This behavior indicates that if a RBF network is constructed using exponential generator functions, the inputs $\mathbf{x}_k$ corresponding to high values of $\|\nabla_{\mathbf{v}_j} h_{j,k}\|^2$ have no significant effect on the response of the radial basis function located at the prototype $\mathbf{v}_j$. As a result, the update of this prototype due to $\mathbf{x}_k$ does not produce significant variations in the input of the upper associative network that implements the desired input-output mapping. Figures 4 and 5 also indicate that the blind spot shrinks as the value of β increases while $\|\nabla_{\mathbf{v}_j} h_{j,k}\|^2$ reaches higher values. Decreasing the value of β expands the blind spot but $\|\nabla_{\mathbf{v}_j} h_{j,k}\|^2$ reaches lower values. In other words, the selection of the value of β in practice involves a trade-off similar to that associated with the selection of the free parameter δ when the radial basis functions are formed by linear generator functions.

8   Handwritten Digit Recognition

8.1   The NIST Databases

Reformulated RBF neural networks were tested and compared with competing techniques on a large-scale handwritten digit recognition problem. The objective of a classifier in this application is the recognition of the digit represented by a binary image of a handwritten numeral. Recognition of handwritten digits is the key component of automated systems developed for a great variety of real-world applications, including mail sorting and check processing. Automated recognition of handwritten digits is not a trivial task due to the high variance of handwritten digits caused by different writing styles, pens, etc. Thus, the development of a reliable system for handwritten digit recognition requires large databases containing a great variety of samples. Such a collection of handwritten digits is contained in the *NIST Special Databases 3*, which contain about 120000 isolated binary digits that have been extracted from sample forms. These

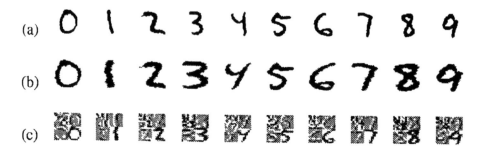

Figure 6. Digits from the NIST Databases: (a) original binary images, (b) 32×32 binary images after one stage of preprocessing (slant and size normalization), and (c) 16 × 16 images of the digits after two stages of preprocessing (slant and size normalization followed by wavelet decomposition).

digits were handwritten by about 2100 field representatives of the United States Census Bureau. The isolated digits were scanned to produce binary images of size 40 × 60 pixels, which are centered in a 128 × 128 box. Figure 6(a) shows some sample digits from 0 to 9 from the NIST databases used in these experiments. The data set was partitioned in three subsets as follows: 58646 digits were used for training, 30367 digits were used for testing, and the remaining 30727 digits constituted the validation set.

8.2   Data Preprocessing

The raw data from the NIST databases were preprocessed in order to reduce the variance of the images that is not relevant to classification. The first stage of the preprocessing scheme produced a slant and size normalized version of each digit. The slant of each digit was found by first determining its center of gravity, which defines an upper and lower half of it. The centers of gravity of each half were subsequently computed and provided an estimate of the vertical main axis of the digit. This axis was then made exactly vertical using a horizontal shear transformation. In the next step, the minimal bounding box was determined and the digit was scaled into a 32 × 32 box. This scaling may slightly distort the aspect ratio of the digits by centering, if necessary, the digits in the box. Figure 6(b) shows the same digits shown in Figure 6(a) after slant and size normalization.

The second preprocessing stage involved a 4-level wavelet decomposition of the 32×32 digit representation produced by the first preprocessing stage. Each decomposition level includes the application of a 2-D Haar wavelet filter in the decomposed image, followed by downsampling by a factor of 2 along the horizontal and vertical directions. Because of downsampling, each decomposition level produces four subbands of lower resolution, namely a subband that carries background information (containing the low-low frequency components of the original subband), two subbands that carry horizontal and vertical details (containing low-high and high-low frequency components of the original subband), and a subband that carries diagonal details (containing the high-high frequency components of the original subband). As a result, the 4-level decomposition of the original 32×32 image produced three subbands of sizes 16×16, 8×8, and 4×4, and four subbands of size 2×2. The 32×32 image produced by wavelet decomposition was subsequently reduced to an image of size 16×16 by representing each 2×2 window by the average of the four pixels contained in it. This step reduces the amount of data by 3/4 and has a smoothing effect that suppresses the noise present in the 32×32 image [1]. Figure 6(c) shows the images representing the digits shown in Figures 6(a) and 6(b), resulting after the second preprocessing stage described above.

8.3   Classification Tools for NIST Digits

This section begins with a brief description of the variants of the *k-nearest neighbor* (*k*-NN) classifier used for benchmarking the performance of the neural networks tested in the experiments and also outlines the procedures used for classifying the digits from the NIST databases using neural networks. These procedures involve the formation of the desired input-output mapping and the strategies used to recognize the NIST digits by interpreting the responses of the trained neural networks to the input samples.

The *k*-NN classifier uses feature vectors from the training set as a reference to classify examples from the testing set. Given an input example from the testing set, the *k*-NN classifier computes its Euclidean distance from all the examples included in the training set. The *k*-NN classifier can be implemented to classify all input examples (no rejections allowed) or to selectively reject some ambiguous examples. The k-NN classifier can be implemented using two alternative classification strategies: According to the first and most frequently used classification strategy, each of the k closest training examples to the input example has a vote with a weight equal to 1. According to the second classification strategy, the ith closest training example to the input example has a vote with weight $1/i$; that is, the weight of the closest example is 1, the weight of the second closest example is $1/2$, etc. When no rejections are allowed, the class that receives the largest sum of votes wins the competition and the input example is assigned the corresponding label. The input example is *recognized* if the label assigned by the classifier and the actual label are identical or *substituted* if the assigned and actual labels are different. When rejections are allowed and the k closest training examples to the input example have votes equal to 1, the example is *rejected* if the largest sum of votes is less than k. Otherwise, the input example is classified according to the strategy described above.

The reformulated RBF neural networks and feed-forward neural networks (FFNNs) tested in these experiments consisted of $256 = 16 \times 16$ inputs and 10 linear output units, each representing a digit from 0 to 9. The inputs of the networks were normalized to the interval $[0, 1]$. The learning rate in all these experiments was $\eta = 0.1$. The networks were trained to respond with $y_{i,k} = 1$ and $y_{j,k} = 0$, $\forall j \neq i$, when presented with an input vector $\mathbf{x}_k \in \mathcal{X}$ corresponding to the digit represented by the ith output unit. The assignment of input vectors to classes was based on a winner-takes-all strategy. More specifically, each input vector was assigned to the class represented by the output unit of the trained RBF neural network with the maximum response. In an attempt to improve the reliability of the neural-network-based classifiers, label assignment was also implemented by employing an alternative scheme that allows the rejection of some ambiguous digits according to the strategy described below: Suppose one of the trained networks is presented with an input vector $\mathbf{x}$ representing a digit from the testing or the validation set and let $\hat{y}_i$, $1 \leq i \leq n_o$, be the responses of its output units. Let $\hat{y}^{(1)} = \hat{y}_{i_1}$ be the maximum among all responses of the output units, that is, $\hat{y}^{(1)} = \hat{y}_{i_1} = \max_{i \in \mathcal{I}_1}\{\hat{y}_i\}$, with $\mathcal{I}_1 = \{1, 2, \ldots, n_o\}$. Let $\hat{y}^{(2)} = \hat{y}_{i_2}$ be the maximum among the responses of the rest of the output units, that is, $\hat{y}^{(2)} = \hat{y}_{i_2} = \max_{i \in \mathcal{I}_2}\{\hat{y}_i\}$, with $\mathcal{I}_2 = \mathcal{I}_1 - \{i_1\}$. The simplest classification scheme would be to assign the digit represented by **x** to the i_1th class, which implies that none of the digits would be rejected by the network. Nevertheless, the reliability of this assignment can be improved by comparing the responses $\hat{y}^{(1)}$ and $\hat{y}^{(2)}$ of the two output units that claim the digit for their corresponding classes. If the responses $\hat{y}^{(1)}$ and $\hat{y}^{(2)}$ are sufficiently close, then the digit represented by **x** probably lies in a region of the input space where the classes represented by the i_1th and i_2th output units overlap. This indicates that the reliability of classification can be improved by rejecting this digit. This rejection strategy can be implemented by comparing the difference $\Delta\hat{y} = \hat{y}^{(1)} - \hat{y}^{(2)}$ with a *rejection parameter* $r \geq 0$. The digit corresponding to **x** is *accepted* if $\Delta\hat{y} \geq r$ and *rejected* otherwise. An accepted digit is *recognized* if the output unit with the maximum response represents the desired class and *substituted* otherwise. The rejection rate depends on the selection of the rejection parameter $r \geq 0$. If $r = 0$, then the digit corresponding to **x** is accepted if $\Delta\hat{y} = \hat{y}^{(1)} - \hat{y}^{(2)} \geq 0$, which is by definition true. This implies that none of the input digits is rejected by the network if $r = 0$. The rejection rate increases as the value of the rejection parameter increases above 0.

8.4   Role of the Prototypes in Gradient Descent Learning

RBF neural networks are often trained to implement the desired input-output mapping by updating the output weights, that is, the weights that connect the radial basis functions and the output units, in order to minimize the output error. The radial basis functions are centered at a fixed set of prototypes that define a partition of the input space. In contrast, the gradient descent algorithm presented in Section 6 updates the prototypes representing the centers of the radial basis functions together with the output weights every time training examples are presented to the network. This set of experiments investigated the importance of updating the prototypes during the learning process in order to implement the desired input-output mapping. The reformulated RBF neural networks tested in these experiments contained $c = 256$ radial basis functions obtained in terms of the generator function $g_0(x) = 1 + \delta\, x$, with $g(x) = (g_0(x))^{\frac{1}{1-m}}$, $m = 3$ and $\delta = 10$. In all these experiments the prototypes of the

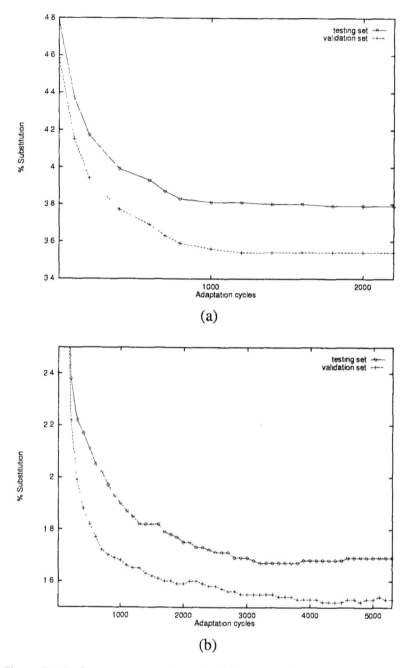

(a)

(b)

Figure 7. Performance of a reformulated RBF neural network tested on the testing and validation sets during its training. The network was trained (a) by updating only its output weights, and (b) by updating its output weights and prototypes.

RBF neural networks were determined by employing the initialization scheme described in Section 6, which involves prototype splitting followed by the c-means algorithm. Figure 7 summarizes the performance of the networks trained in these experiments at different stages of the learning process. Figure 7(a) shows the percentage of digits from the testing and validation sets substituted by the RBF network trained by updating only the output weights while keeping the prototypes fixed during the learning process. Figure 7(b) shows the percentage of digits from the testing and validation sets substituted by the reformulated RBF neural network trained by updating the prototypes and the output weights according to the sequential gradient descent learning algorithm presented in Section 6. In both cases, the percentage of substituted digits decreased with some fluctuations during the initial adaptation cycles and remained almost constant after a certain number of adaptation cycles. When the prototypes were fixed during learning, the percentage of substituted digits from both testing and validation sets remained almost constant after 1000 adaptation cycles. In contrast, the percentage of substituted digits decreased after 1000 adaptation cycles and remained almost constant after 3000 adaptation cycles when the prototypes were updated together with the output weights using gradient descent. In this case, the percentage of substituted digits reached 1.69% on the testing set and 1.53% on the validation set. This outcome can be compared with the substitution of 3.79% of the digits from the testing set and 3.54% of the digits from the validation set produced when the prototypes remained fixed during learning. This experimental outcome verifies that the performance of RBF neural networks can be significantly improved by updating all their free parameters during learning according to the training set, including the prototypes that represent the centers of the radial basis functions in the input space.

8.5 Effect of the Number of Radial Basis Functions

This set of experiments evaluated the performance on the testing and validation sets formed from the NIST data of various reformulated RBF neural networks at different stages of their training. The reformulated RBF neural networks contained $c = 64$, $c = 128$, $c = 256$, and $c = 512$ radial basis functions obtained in terms of the generator func-

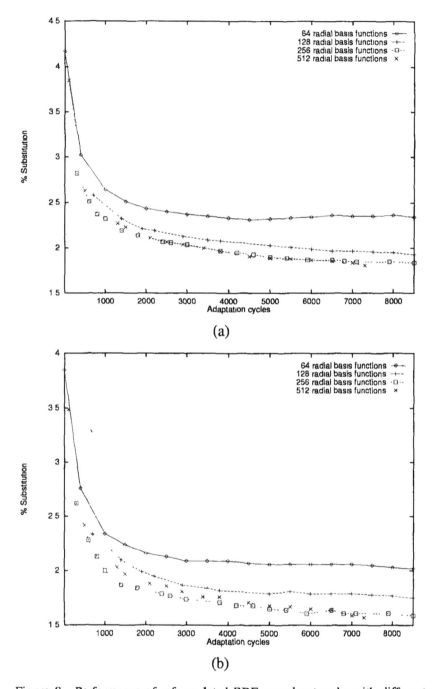

Figure 8. Performance of reformulated RBF neural networks with different numbers of radial basis functions during their training. The substitution rate was computed (a) on the testing set, and (b) on the validation set.

tion $g_0(x) = 1 + \delta x$ as $\phi(x) = g(x^2)$, with $g(x) = (g_0(x))^{\frac{1}{1-m}}$, $m = 3$ and $\delta = 10$. All networks were trained using the sequential gradient descent algorithm described in Section 6. The initial prototypes were computed using the initialization scheme involving prototype splitting. Figures 8(a) and 8(b) plot the percentage of digits from the testing and validation sets, respectively, that were substituted by all four reformulated RBF neural networks as a function of the number of adaptation cycles. Regardless of the number of radial basis functions contained by the reformulated RBF neural networks, their performance on both testing and validation sets improved as the number of adaptation cycles increased. The improvement of the performance was significant during the initial adaptation cycles, which is consistent with the behavior and convergence properties of the gradient descent algorithm used for training. Figures 8(a) and 8(b) also indicate that the number of radial basis functions had a rather significant effect on the performance of reformulated RBF neural networks. The performance of reformulated RBF neural networks on both testing and validation sets improved as the number of radial basis functions increased from $c = 64$ to $c = 128$. The best performance on both sets was achieved by the reformulated RBF neural networks containing $c = 256$ and $c = 512$ radial basis functions. It must be noted that there are some remarkable differences in the performance of these two networks on the testing and validation sets. According to Figure 8(a), the reformulated RBF neural networks with $c = 256$ and $c = 512$ radial basis functions substituted almost the same percentage of digits from the testing set after 1000 adaptation cycles. However, the network with $c = 512$ radial basis functions performed slightly better on the testing set than that containing $c = 256$ radial basis functions when the training continued beyond 7000 adaptation cycles. According to Figure 8(b), the reformulated RBF network with $c = 256$ radial basis functions outperformed consistently the network containing $c = 512$ radial basis functions on the validation set for the first 6000 adaptation cycles. However, the reformulated RBF network with $c = 512$ radial basis functions substituted a smaller percentage of digits from the validation set than the network with $c = 256$ radial basis functions when the training continued beyond 7000 adaptation cycles.

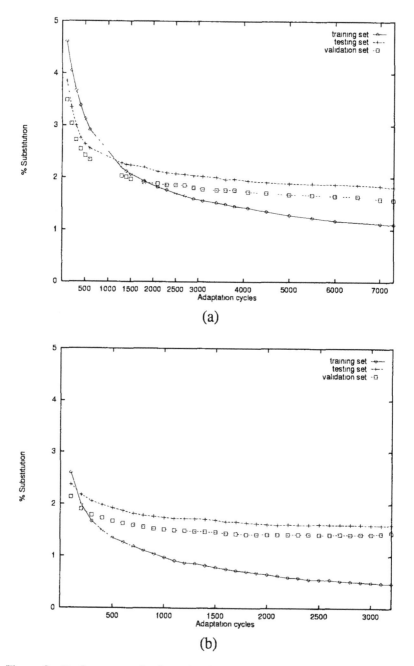

(a)

(b)

Figure 9. Performance of reformulated RBF neural networks with 512 radial basis functions during their training. The substitution rates were computed on the training, testing, and validation sets when gradient descent training was initialized (a) randomly, and (b) by prototype splitting.

8.6   Effect of the Initialization of Gradient Descent Learning

This set of experiments evaluated the effect of the initialization of the supervised learning on the performance of reformulated RBF neural networks trained by gradient descent. The reformulated RBF neural network tested in these experiments contained $c = 512$ radial basis functions constructed as $\phi(x) = g(x^2)$, with $g(x) = (g_0(x))^{\frac{1}{1-m}}$, $g_0(x) = 1 + \delta x$, $m = 3$, and $\delta = 10$. The network was trained by the sequential gradient descent algorithm described in Section 6. Figures 9(a) and 9(b) show the percentage of digits from the training, testing, and validation sets substituted during the training process when gradient descent learning was initialized by randomly selecting the prototypes and by prototype splitting, respectively. When the initial prototypes were determined by prototype splitting, the percentage of substituted digits from the training set decreased below 1% after 1000 adaptation cycles and reached values below 0.5% after 3000 adaptation cycles. In contrast, the percentage of substituted digits from the training set decreased much slower and never reached values below 1% when the initial prototypes were produced by a random number generator. When the initial prototypes were initialized by prototype splitting, the percentage of substituted digits from the testing and validation sets decreased to values around 1.5% after the first 1000 adaptation cycles and changed very slightly as the training progressed. When the supervised training was initialized randomly, the percentage of substituted digits from the testing and validation sets decreased much slower during training and reached values higher than those shown in Figure 9(b) even after 7000 adaptation cycles. This experimental outcome indicates that initializing gradient descent learning by prototype splitting improves the convergence rate of gradient descent learning and leads to trained networks that achieve superior performance.

8.7   Benchmarking Reformulated RBF Neural Networks

The last set of experiments compared the performance of reformulated RBF neural networks trained by gradient descent with that of FFNNs

Table 1. Substitution rates on the testing set (S_{test}) and the validation set (S_{val}) produced for different values of k by two variants of the k-NN classifier when no rejections were allowed.

	k-NN classifier (equal vote weights)		k-NN classifier (unequal vote weights)	
k	S_{test}	S_{val}	S_{test}	S_{val}
2	2.351	2.128	2.210	2.018
4	2.025	1.917	2.029	1.852
8	2.055	1.959	1.969	1.836
16	2.259	2.099	1.897	1.832
32	2.496	2.353	1.923	1.875
64	2.869	2.724	2.002	1.949

Table 2. Substitution rates on the testing set (S_{test}) and the validation set (S_{val}) produced by different neural-network-based classifiers when no rejections were allowed. FFNNs and reformulated RBF neural networks (RBFNNs) were trained with different numbers c of hidden units by gradient descent. The training of reformulated RBF neural networks was initialized randomly and by prototype splitting.

	FFNN		RBFNN		RBFNN (+ splitting)	
c	S_{test}	S_{val}	S_{test}	S_{val}	S_{test}	S_{val}
64	2.63	2.40	2.31	2.02	2.24	2.03
128	1.82	1.74	1.92	1.75	1.93	1.75
256	1.81	1.59	1.84	1.59	1.62	1.47
512	1.89	1.63	1.81	1.57	1.60	1.41

with sigmoidal hidden units and the k-NN classifier. The success rate was first measured when these classifiers were required to assign class labels to all input vectors corresponding to the digits from the testing and validation sets. Table 1 summarizes the substitution rates produced on the testing and validation sets by the two variants of the k-NN algorithm used for recognition when no rejections were allowed. The values of k were powers of two varying from 2 to 64. When each of the k closest training examples voted with weight 1, the smallest substitution rate was recorded for $k = 4$. When each of the k closest training examples voted according to their distance from the input example, the smallest substitution rate was recorded for $k = 16$. In this case, increasing the value of k up to 16 decreased the substitution rate. This can be attributed to the fact that the votes of all k training examples were weighted with values that decreased from 1 to $1/k$, which reduced the contribution of the most distant among the k training examples. This weighting strategy improved the performance of the k-NN classifier, as indicated by Table 1. When no rejections were allowed, the performance of both variants of the k-NN classifier was inferior to that of the neural networks tested in these experiments. This is clearly indicated by Table 2, which summarizes the substitution rates produced on the testing and validation sets by FFNNs and reformulated RBF neural networks. The number of hidden units varied in these experiments from 64 to 512. The sets of prototypes used for initializing the supervised training of reformulated RBF neural networks were produced by a random number generator and by the prototype splitting procedure outlined in Section 6. The performance of the trained FFNNs on both testing and validation sets improved consistently as the number of hidden units increased from 64 to 256 but degraded when the number of hidden units increased from 256 to 512. In contrast, the performance of reformulated RBF neural networks on both testing and validation sets improved consistently as the number of radial basis function units increased from $c = 64$ to $c = 512$. Both reformulated RBF neural networks trained with $c = 512$ radial basis functions outperformed the best FFNN. Moreover, the performance of the best FFNN was inferior to that of the reformulated RBF neural network trained with $c = 256$ radial basis functions using the initialization scheme employing prototype splitting. The best overall performance among all classifiers evaluated in this set of experiments was achieved by the reformulated RBF neural network trained with $c = 512$ radial basis functions by gradient descent initialized by prototype splitting.

The success rate of the k-NN classifier and the neural-network-based classifiers was also measured when these classifiers were allowed to reject some ambiguous digits in order to improve their reliability. The k-NN classifier was implemented in these experiments by assigning votes equal to 1 to the k closest training examples. This variant of the k-NN classifier does not reject any digit if $k = 1$. The percentage of digits rejected by this variant of the k-NN classifier increases as the value of k increases. The rejection of digits by the FFNN and reformulated RBF neural networks was based on the strategy outlined above. According to this strategy, the percentage of the rejected digits increases as the rejec-

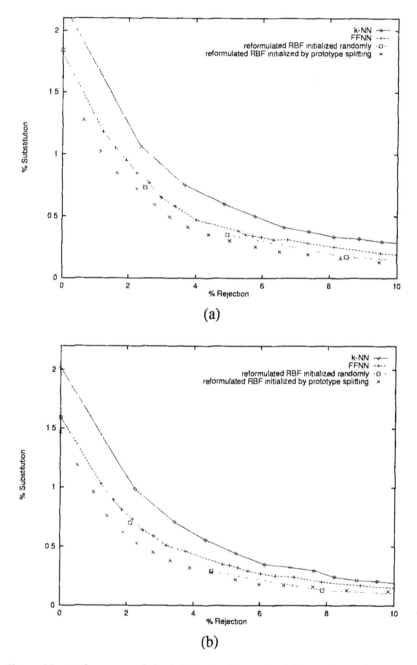

Figure 10. Performance of the k-NN classifier, a feed-forward neural network and two reformulated RBF neural networks tested on the NIST digits. The substitution rate is plotted versus the rejection rate (a) on the testing set, and (b) on the validation set.

tion parameter r increases above 0. For $r = 0$, no digits are rejected and classification is based on a winner-takes-all strategy. Figure 10 plots the percentage of digits from the testing and validation sets substituted at different rejection rates by the k-NN classifier, an FFNN with 256 hidden units, and two reformulated RBF neural networks with 256 radial basis functions. Both RBF neural networks were trained by the sequential gradient descent algorithm presented in Section 6. The supervised learning process was initialized in one case by randomly generating the initial set of prototypes and in the other case by determining the initial set of prototypes using prototype splitting. The training of all neural networks tested was terminated based on their performance on the testing set. When no rejections were allowed, all neural networks tested in these experiments performed better than various classification schemes tested on the same data set [10], none of which exceeded the recognition rate of 97.5%. In this case, all neural networks outperformed the k-NN classifier, which classified correctly 97.79% of the digits from the testing set and 97.98% of the digits from the validation set. When no rejections were allowed, the best performance was achieved by the reformulated RBF neural network whose training was initialized by the prototype splitting procedure outlined in Section 6. This network classified correctly 98.38% of the digits from the testing set and 98.53% of the digits from the validation set. According to Figure 10, the percentage of digits from the testing and validation sets substituted by all classifiers tested in these experiments decreased as the rejection rate increased. This experimental outcome verifies that the strategy employed for rejecting ambiguous digits based on the outputs of the trained neural networks is a simple and effective way of dealing with uncertainty. Regardless of the rejection rate, all three neural networks tested in these experiments outperformed the k-NN classifier, which substituted the largest percentage of digits from both testing and validation sets. The performance of the reformulated RBF neural network whose training was initialized by randomly generating the prototypes was close to that of the FFNN. In fact, the FFNN performed better at low rejection rates while this reformulated RBF neural network outperformed the FFNN at high rejection rates. The reformulated RBF neural network initialized by the prototype splitting procedure outlined in Section 6 performed consistently better on the testing and validation sets than the FFNN. The same RBF network outperformed the reformulated RBF neural network initialized randomly on the testing set and on the validation set for low rejection rates. However, the two reformulated RBF neural networks achieved the same digit recognition rates on the **validation set as the rejection rate** increased. Among the three networks tested, the best overall performance was achieved by the reformulated RBF neural network whose training was initialized using prototype splitting.

9   Conclusions

This chapter presented an axiomatic approach for reformulating RBF neural networks trained by gradient descent. According to this approach, the development of admissible RBF models reduces to the selection of admissible generator functions that determine the form and properties of the radial basis functions. The reformulated RBF neural networks generated by linear and exponential generator functions can be trained by gradient descent and perform considerably better than conventional RBF neural networks. The criteria proposed for selecting generator functions indicated that linear generator functions have certain advantages over exponential generator functions, especially when reformulated RBF neural networks are trained by gradient descent. Given that exponential generator functions lead to Gaussian radial basis functions, the comparison of linear and exponential generator functions indicated that Gaussian radial basis functions are not the only, and perhaps not the best, choice for constructing RBF neural models. Reformulated RBF neural networks were originally constructed using linear functions of the form $g_0(x) = x + \gamma^2$, which lead to a family of radial basis functions that includes inverse multiquadratic radial basis functions [13], [20], [21]. Subsequent studies, including that presented in the chapter, indicated that linear functions of the form $g_0(x) = 1 + \delta x$ facilitate the training and improve the performance of reformulated RBF neural networks [17].

The experimental evaluation of reformulated RBF neural networks presented in this chapter showed that the association of RBF neural networks with erratic behavior and poor performance is unfair to this powerful neural architecture. The experimental results also indicated that the disadvantages often associated with RBF neural networks can only be attributed to the learning schemes used for their training and not to the models themselves. If the learning scheme used to train RBF neural networks decouples the determination of the prototypes and the updates of the output weights, then the prototypes are simply determined to satisfy the optimization criterion behind the unsupervised algorithm employed. Nevertheless, the satisfaction of this criterion does not necessarily guarantee that the partition of the input space by the prototypes facilitates the implementation of the desired input-output mapping. The simple reason for this is that the training set does not participate in the formation of the prototypes. In contrast, the update of the prototypes during the learning process produces a partition of the input space that is specifically designed to facilitate the input-output mapping. In effect, this partition leads to trained reformulated RBF neural networks that are strong competitors to other popular neural models, including feed-forward neural networks with sigmoidal hidden units.

The results of the experiments on the NIST digits verified that reformulated RBF neural networks trained by gradient descent are strong competitors to classical classification techniques, such as the k-NN, and alternative neural models, such as FFNNs. The digit recognition rates achieved by reformulated RBF neural networks were consistently higher than those of feed-forward neural networks. The classification accuracy of reformulated RBF neural networks was also found to be superior to that of the k-NN classifier. In fact, the k-NN classifier was outperformed by all neural networks tested in these experiments. Moreover, the k-NN classifier was computationally more demanding than all the trained neural networks, which classified examples much faster than the k-NN classifier. The time required by the k-NN to classify an example increased with the problem size (number of examples in the training set), which had absolutely no effect on the classification of digits by the trained neural networks. The experiments on the NIST digits also indicated that the reliability and classification accuracy of trained neural networks can be improved by a recall strategy that allows the rejection of some ambiguous digits.

The experiments indicated that the performance of reformulated RBF neural networks improves when their supervised training by gradient descent is initialized by using an effective unsupervised procedure to determine the initial set of prototypes from the input vectors included in the training set. An alternative to employing the variation of the c-means algorithm employed in these experiments would be the use of unsupervised algorithms that are not significantly affected by their initialization. The search for such codebook design techniques led to soft clustering [2], [11], [14], [18], [19] and soft learning vector quantization algorithms [12], [15], [16], [18], [19], [24], [27], [35], [41]. Unlike crisp clustering and vector quantization techniques, these algorithms form the prototypes on the basis of *soft* instead of crisp decisions. As a result, this strategy reduces significantly the effect of the initial set of prototypes on the partition of the input vectors produced by such algorithms. The use of soft clustering and LVQ algorithms for initializing the training of reformulated RBF neural networks is a particularly promising approach currently under investigation. Such an initialization approach is strongly supported by recent developments in unsupervised competitive learning, which indicated that the same generator functions used for constructing reformulated RBF neural networks can also be used to generate soft LVQ and clustering algorithms [19], [20], [22].

The generator function can be seen as the concept that establishes a direct relationship between reformulated RBF models and soft LVQ algorithms [20]. This relationship makes reformulated RBF models potential targets of the search for architectures inherently capable of merging neural modeling with fuzzy-theoretic concepts, a problem that attracted considerable attention recently [39]. In this context, a problem worth investigating is the ability of reformulated RBF neural networks to detect the presence of uncertainty in the training set and quantify the existing uncertainty by approximating any membership profile arbitrarily well from sample data.

References

[1] Behnke, S. and Karayiannis, N.B. (1998), "Competitive neural trees for pattern classification," *IEEE Transactions on Neural Networks,* vol. 9, no. 6, pp. 1352-1369.

[2] Bezdek, J.C. (1981), *Pattern Recognition with Fuzzy Objective Function Algorithms,* Plenum, New York, NY.

[3] Broomhead, D.S. and Lowe, D. (1988), "Multivariable functional

interpolation and adaptive networks," *Complex Systems*, vol. 2, pp. 321-355.

[4] Cha, I. and Kassam, S.A. (1995), "Interference cancellation using radial basis function networks," *Signal Processing,* vol. 47, pp. 247-268.

[5] Chen, S., Cowan, C.F.N., and Grant, P.M. (1991), "Orthogonal least squares learning algorithm for radial basis function networks," *IEEE Transactions on Neural Networks,* vol. 2, no. 2, pp. 302-309.

[6] Chen, S., Gibson, G.J., Cowan, C.F.N., and Grant, P.M. (1991), "Reconstruction of binary signals using an adaptive radial-basis-function equalizer," *Signal Processing,* vol. 22, pp. 77-93.

[7] Cybenko, G. (1989), "Approximation by superpositions of a sigmoidal function," *Mathematics of Control, Signals, and Systems,* vol. 2, pp. 303-314.

[8] Gersho, A. and Gray, R.M. (1992), *Vector Quantization and Signal Compression,* Kluwer Academic, Boston, MA.

[9] Gray, R.M. (1984), "Vector quantization," *IEEE ASSP Magazine,* vol. 1, pp. 4-29.

[10] Grother, P.J. and Candela, G.T. (1993), "Comparison of handprinted digit classifiers," *Technical Report NISTIR 5209,* National Institute of Standards and Technology, Gaithersburg, MD.

[11] Karayiannis, N.B. (1996), "Generalized fuzzy *c*-means algorithms," *Proceedings of Fifth International Conference on Fuzzy Systems,* New Orleans, LA, pp. 1036-1042.

[12] Karayiannis, N.B. (1997), "Entropy constrained learning vector quantization algorithms and their application in image compression," *SPIE Proceedings vol. 3030: Applications of Artificial Neural Networks in Image Processing II,* San Jose, CA, pp. 2-13.

[13] Karayiannis, N.B. (1997), "Gradient descent learning of radial basis neural networks," *Proceedings of 1997 IEEE International Conference on Neural Networks,* Houston, TX, pp. 1815-1820.

[14] Karayiannis, N.B. (1997), "Fuzzy partition entropies and entropy constrained clustering algorithms," *Journal of Intelligent & Fuzzy Systems,* vol. 5, no. 2, pp. 103-111.

[15] Karayiannis, N.B. (1997), "Learning vector quantization: A review," *International Journal of Smart Engineering System Design,* vol. 1, pp. 33-58.

[16] Karayiannis, N.B. (1997), "A methodology for constructing fuzzy algorithms for learning vector quantization," *IEEE Transactions on Neural Networks,* vol. 8, no. 3, pp. 505-518.

[17] Karayiannis, N.B. (1998), "Learning algorithms for reformulated radial basis neural networks," *Proceedings of 1998 International Joint Conference on Neural Networks,* Anchorage, AK, pp. 2230-2235.

[18] Karayiannis, N.B. (1998), "Ordered weighted learning vector quantization and clustering algorithms," *Proceedings of 1998 International Conference on Fuzzy Systems,* Anchorage, AK, pp. 1388-1393.

[19] Karayiannis, N.B. (1998), "Soft learning vector quantization and clustering algorithms based in reformulation," *Proceedings of 1998 International Conference on Fuzzy Systems,* Anchorage, AK, pp. 1441-1446.

[20] Karayiannis, N.B. (1999), "Reformulating learning vector quantization and radial basis neural networks," *Fundamenta Informaticae,* vol. 37, pp. 137-175.

[21] Karayiannis, N.B. (1999), "Reformulated radial basis neural networks trained by gradient descent," *IEEE Transactions on Neural Networks,* vol. 10, no. 3, pp. 657-671.

[22] Karayiannis, N.B. (1999), "An axiomatic approach to soft learning vector quantization and clustering," *IEEE Transactions on Neural Networks,* vol. 10, no. 5, pp. 1153-1165.

[23] Karayiannis, N.B. and Bezdek, J.C. (1997), "An integrated approach to fuzzy learning vector quantization and fuzzy *c*-means

clustering," *IEEE Transactions on Fuzzy Systems*, vol. 5, no. 4, pp. 622-628.

[24] Karayiannis, N.B., Bezdek, J.C., Pal, N.R., Hathaway, R.J., and Pai, P.-I (1996), "Repairs to GLVQ: A new family of competitive learning schemes," *IEEE Transactions on Neural Networks*, vol. 7, no. 5, pp. 1062-1071.

[25] Karayiannis, N.B. and Mi, W. (1997), "Growing radial basis neural networks: Merging supervised and unsupervised learning with network growth techniques," *IEEE Transactions on Neural Networks*, vol. 8, no. 6, pp. 1492-1506.

[26] Karayiannis, N.B. and Pai, P.-I (1995), "Fuzzy vector quantization algorithms and their application in image compression," *IEEE Transactions on Image Processing*, vol. 4, no. 9, pp. 1193-1201.

[27] Karayiannis, N.B. and Pai, P.-I (1996), "Fuzzy algorithms for learning vector quantization," *IEEE Transactions on Neural Networks*, vol. 7, no. 5, pp. 1196-1211.

[28] Karayiannis, N.B. and Venetsanopoulos, A.N. (1993), *Artificial Neural Networks: Learning Algorithms, Performance Evaluation, and Applications*, Kluwer Academic, Boston, MA.

[29] Kohonen, T. (1989), *Self-Organization and Associative Memory*, 3rd Edition, Springer-Verlag, Berlin.

[30] Kohonen, T. (1990), "The self-organizing map," *Proceeding of the IEEE*, vol. 78, no. 9, pp. 1464-1480.

[31] Linde, Y., Buzo, A., and Gray, R.M. (1980), "An algorithm for vector quantization design," *IEEE Transactions on Communications*, vol. 28, no. 1, pp. 84-95.

[32] Lippmann, R.P. (1989), "Pattern classification using neural networks," *IEEE Communications Magazine*, vol. 27, pp. 47-54.

[33] Micchelli, C.A. (1986), "Interpolation of scattered data: Distance matrices and conditionally positive definite functions," *Constructive Approximation*, vol. 2, pp. 11-22.

[34] Moody, J.E. and Darken, C.J. (1989), "Fast learning in networks of locally-tuned processing units," *Neural Computation*, vol. 1, pp. 281-294.

[35] Pal, N.R., Bezdek, J.C., and Tsao, E.C.-K. (1993), "Generalized clustering networks and Kohonen's self-organizing scheme," *IEEE Transactions on Neural Networks*, vol. 4, no. 4, pp. 549-557.

[36] Park, J. and Sandberg, I.W. (1991), "Universal approximation using radial-basis-function networks," *Neural Computation*, vol. 3, pp. 246-257.

[37] Park, J. and Sandberg, I.W. (1993), "Approximation and radial-basis-function networks," *Neural Computation*, vol. 5, pp. 305-316.

[38] Poggio, T. and Girosi, F. (1990), "Regularization algorithms for learning that are equivalent to multilayer networks," *Science*, vol. 247, pp. 978-982.

[39] Purushothaman, G. and Karayiannis, N.B. (1997), "Quantum Neural Networks (QNNs): Inherently fuzzy feedforward neural networks," *IEEE Transactions on Neural Networks*, vol. 8, no. 3, pp. 679-693.

[40] Roy, A., Govil, S., and Miranda, R. (1997), "A neural-network learning theory and a polynomial time RBF algorithm," *IEEE Transactions on Neural Networks,* vol. 8, no. 6, pp. 1301-1313.

[41] Tsao, E.C.-K., Bezdek, J.C., and Pal, N.R. (1994), "Fuzzy Kohonen clustering networks," *Pattern Recognition*, vol. 27, no. 5, pp. 757-764.

[42] Whitehead, B.A. and Choate, T.D. (1994), "Evolving space-filling curves to distribute radial basis functions over an input space," *IEEE Transactions on Neural Networks,* vol. 5, no. 1, pp. 15-23.

CHAPTER 3

EFFICIENT NEURAL NETWORK-BASED METHODOLOGY FOR THE DESIGN OF MULTIPLE CLASSIFIERS

N. Vassilas

Institute of Informatics and Telecommunications
National Research Center "Demokritos," Ag. Paraskevi, Attiki
Greece

A neural network-based methodology for time and memory efficient supervised or unsupervised classification in heavily demanding applications is presented in this chapter. Significantly increased speed in the design (training) of neural, fuzzy and statistical classifiers as well as in the classification phase is achieved by: (a) using a self-organizing feature map (SOFM) for vector quantization and indexed representation of the input data space; (b) appropriate training set reduction using the SOFM prototypes followed by necessary modifications of the training algorithms (supervised techniques); (c) clustering of neurons on maps instead of clustering the original data (unsupervised techniques); and (d) fast indexed classification. Finally, a demonstration of this methodology involving the design of multiple classifiers is performed on Land-Cover classification of multispectral satellite image data showing increased speed with respect to both training and classification times.

1 Introduction

Within the last decade, advances in space, sensor and computer technology combined with the launch of new sophisticated satellites have made it possible to amass huge amounts of data about the earth and its environment daily [1]. Applications such as environmental monitoring and resource management as well as geological and geophysical data analysis involve processing large amounts of spatial

0-8493-2268-5/2000/$0.00+$.50

data. Often, these data are collected from multiple sources, stored in Geographical Information Systems (GIS), and may include multispectral satellite images (e.g., Landsat TM, SPOT, NOAA AVHRR), grid data (e.g., digital elevation maps, geological/geophysical maps) and point or local measurements (e.g., data from local stations, drill data). Therefore, it is evident that more powerful methodologies are needed for efficient data processing in heavily demanding applications such as the one considered in this paper, namely, satellite image classification.

Recently, several techniques have been proposed for multispectral satellite image classification. These include traditional statistics, neural networks and fuzzy logic and can be divided into two general categories: (a) supervised techniques in which labeled training samples are used for optimizing the design parameters of the classification system [2]-[7], and (b) unsupervised techniques (automatic classification) using a data clustering algorithm [8]-[12].

Although supervised techniques generally perform better in the production of thematic maps (e.g., classification in land-cover categories, geological categories, etc.), unsupervised techniques are mainly used when no training sets are available and constitute a valuable objective alternative as they do not depend on previous knowledge or a photointerpreter's experience. These algorithms first cluster the data according to a similarity criterion, then assign a label to each cluster (usually a grey level or color) that corresponds to a (thematic) category and, finally, substitute each pixel of the original image with the cluster label to which it belongs. The traditional classification scheme using a supervised algorithm or a clustering algorithm is shown in Figure 1.

In a number of recent works [2]-[7], neural network models have successfully been applied for the classification of multispectral satellite images and, more generally, of multisource remotely sensed data. However, for training sets consisting of several thousand patterns belonging to many (often more than ten) categories and large volumes of data, the neural network training and/or classification times reported are quite long, ranging in some cases from a few hours to a few weeks on a conventional computer [4]. The inherent computational complexity in the training, clustering or classification phase of several other algorithms such as the Pal-Majumder's fuzzy classifier [13], [14], the k-nearest neighbors algorithm [15], the various hierarchical clustering procedures [15] and clustering based on scale-space analysis [11], to name a few, also prohibit their use in heavily demanding applications. Therefore, taking into account that additional training and classification trials must usually be performed before selecting a particular classification model, its architecture, and, its design parameters, the need for a methodology for fast model design and classification is evident.

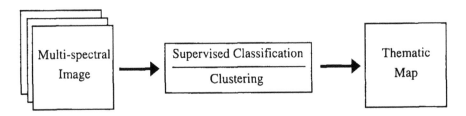

Figure 1. Traditional supervised/unsupervised classification.

In this paper, a methodology based on self-organizing feature maps and indexing techniques is proposed and demonstrated for classifying multispectral satellite images in land-cover categories. The aim is to improve memory requirements for storing the satellite data and at the same time increase training and classification speed without a significant compromise of final performance. This can be accomplished through quantization of the input space, indexed representation of image data, reduction of training (and, possibly, validation) sets, appropriate modification of the training algorithms and, finally, indexed classification.

Results using neural, fuzzy and statistical classifiers show that it is possible to obtain good land-cover maps with the proposed methodology in much shorter times than the classical method of pixel-by-pixel classification. Furthermore, the increased speed achieved allows the design of multiple independent classifiers needed by multimodular systems that combine the decisions of individual classifiers through a voting scheme. One such system that resolves "don't know" cases (i.e., classifiers not in agreement) through local spatial voting is shown to further improve the final result.

2   Proposed Methodology

In this section, a methodology for efficient classification of multi-spectral data is presented with the following advantages:

- memory savings through data quantization,

- increased training speed in supervised algorithms, due to training set size reduction achieved by redundancy removal,

- increased clustering speed in automatic classification, due to the relatively small number of prototypes (quantized data points),

- increased classification speed by using fast indexing techniques.

Although the method can be applied to multisource data without loss of generality, we will restrict the presentation to multispectral satellite images consisting of n bands with $M{\times}N$ pixels each. The image is represented in the n-dimensional euclidean space $\mathbf{R}^n$ by a set of $M{\times}N$ points, whereby the grey level values (intensities) in each band at a particular pixel are interpreted as the coordinates of the corresponding point in $\mathbf{R}^n$ (see Figure 2). In other words, the grey levels of each pixel are stacked into a vector (also called the spectral signature of the pixel) that specifies a point in $\mathbf{R}^n$.

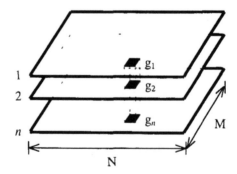

Figure 2. Representation of multispectral images of $M{\times}N$ pixels and n bands. The spectral signature of a pixel is represented by a point $(g_1, g_2, ..., g_n) \in \mathbf{R}^n$ where g_i corresponds to the grey level of the i-th spectral band, $i = 1, 2, ..., n$.

2.1   Data Quantization Using Self-Organizing Maps

The first stage of the proposed methodology involves quantization of the input data space using Kohonen's self-organizing feature maps (SOFM) [16]. Using the euclidean distance metric, the SOFM algorithm performs a Voronoi tessellation of the input space and the asymptotic weights of the network (usually a 1-D or 2-D lattice of neurons) can then be considered as a catalogue of vectors or prototypes, with each such prototype representing all data from its corresponding Voronoi cell.

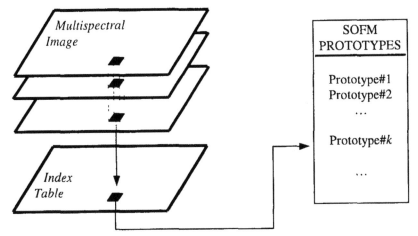

Figure 3.   Representation of multispectral images by the index table and SOFM prototypes. The index table stores pointers from pixels to their nearest prototypes.

Following input data quantization, the next step is to derive an indexing scheme that maps each input sample (pixel) to its corresponding prototype. This is achieved by storing, in a 2-D array of the same dimensions as the original image, a pointer to the pixel's closest prototype. This array will be called the *index table* and, along with the SOFM prototypes, constitutes an indexed (compressed) representation that can be used in place of the original image (see Figure 3). Although SOFM training as well as index table production are performed with off-line computations, it is worth noting that the speed can be significantly increased a) by using the branch-and-bound [17] or partition [5] variants of the nearest neighbor algorithm for sequential implementations, or b) through parallel implementations. Recent works on systolic array implementations of Kohonen's algorithm, which exploit synaptic-level parallelism and allow for fast computations needed in SOFM training and index table construction, can be found in [18], [19].

In general, the larger the number of neurons on the map the better the approximation of the original data space and the smaller the quantization distortion (provided that the map self-organizes). However, from experience, map sizes of up to 16×16 neurons (i.e., 256 prototypes) should suffice in most applications. In the case of large volumes of multispectral data from n bands with 256 grey levels/band, compression ratios of approximately n:1, when 256 prototypes are used, are readily attainable.

2.2 Training Set Reduction and Classification of SOFM Prototypes for Supervised Techniques

In this section, we demonstrate how fast tuning the parameters of supervised algorithms can be performed using the SOFM prototypes. The training phase involves the use of appropriately selected training and, possibly, validation samples of known classification, with the latter being used to avoid *overtraining* [20]. In satellite image classification applications, these data sets are usually composed of several thousands of pixels and, along with the complexity of the classification task (i.e., the number of categories as well as the optimal shapes of class boundaries), are responsible for the long training times observed. Therefore, it is plausible to seek a reduction of these sets through quantization, preserving at the same time most of the information contained in the original sets. There are two main reasons for this: a) it removes redundancy from the training set, and b) such a reduction speeds up the validation set performance evaluation computed at regular intervals during the training phase.

Using the proposed methodology, a reduction of the training and validation sets can be achieved as follows. First, both sets are quantized by substitution of each sample (spectral signature) with its closest SOFM prototype. In general, the number of prototypes is much smaller than the size of either data set, therefore leading to the existence of many duplicated quantized samples (they fall in the same Voronoi cell). Second, for each class label, we partition the data sets in groups of identical samples and compute the multiplicity of each group. The reduced sets will have as many different samples as the number of groups under each label with each sample followed by its multiplicity in the group. These multiplicity counts are used in order to preserve the between- and within-class relative frequencies needed to specify optimal boundary placement in overlapping regions (note that identical samples belonging to different classes will both exist in the reduced set). As will be shown in Section 3, simple algorithmic modifications, taking into account sample multiplicities, allow for fast training of supervised models with reduced training and validation sets.

Finally, in the next stage of the proposed methodology the so trained, supervised models are used to classify the weight vectors associated with the neurons of the map (i.e., the catalogue of SOFM prototypes) rather than the original multispectral data. The result obtained is a catalogue of labels (e.g., grey levels or colors) following the same order as the SOFM prototypes (see Figure 4).

Figure 4.   Creation of the catalogue of labels by supervised models based on the reduced training and validation sets.

2.3   Fast Clustering and Labeling of SOFM Prototypes for Unsupervised Techniques

Typically, automatic classification involves clustering of the data space followed by label assignment. However, due to the large number of data points (up to $M{\times}N$ different spectral signatures), clustering performed on the original image data is inefficient in terms of both memory and time.

In the proposed methodology clustering is performed on the neurons of the map (i.e., the catalogue of SOFM prototypes), thus achieving an increased speed of orders of magnitude, allowing the use of even the most computationally demanding algorithms such as hierarchical algorithms [15].

Following clustering, the next step is the assignment of arbitrary labels to each cluster. These clusters, along with their labels, will represent the automatic classification categories (see Figure 5).

Figure 5. Creation of the catalogue of labels by unsupervised models based on fast clustering and arbitrary label assignment.

2.4   Efficient Indexed Classification

The traditional pixel by pixel classification using supervised or unsupervised techniques requires computational time proportional to the original image dimensions (see Figure 1). The classification of SOFM prototypes and production of the catalogue of labels allows for fast indexed classification by avoiding expensive computations. The final result (thematic map) is now obtained by following the pointers of the index table and accessing the corresponding labels as shown in Figure 6. For large satellite images, an increase in speed of two or three orders of magnitude is possible at this stage.

3   Modifications of Supervised Algorithms

Next, we present the neural, fuzzy and statistical supervised classifiers used in this work as well as the modifications needed in order to take advantage of reduced training and validation sets. In particular, we show how to take into account sample multiplicities when updating weights of neural networks and how to accommodate these multiplicities into fuzzy and statistical classifiers in order to accelerate their computational performance.

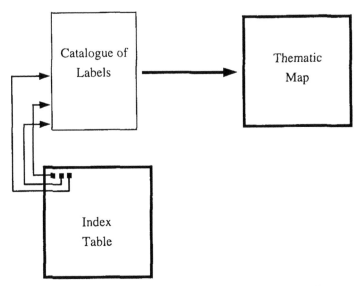

Figure 6. Fast indexed classification using the index table and catalogue of labels.

As far as neural networks are concerned, the goal is for the weight updates to be equivalent to those that correspond to the unreduced but quantized training sets. Hence, under these modifications and assuming that the complexity of the classification task is not significantly affected by the quantization process, training time is reduced approximately in proportion to the ratio between the original (redundant) training set size and the number of prototypes (provided that most of the prototypes exist in the reduced set with few belonging to different classes).

3.1   Classification Using the BP Algorithm

The first classifier used is a multi-layered feedforward network trained with the back propagation (BP) algorithm [21]. In order to speed up the convergence to a local minimum of the error surface, by allowing for relatively large adaptation gain (learning rate) parameters, the *on-line* back propagation version [22] is used in the simulations. In on-line BP, the weights are updated following each input presentation, whereby input patterns are provided in random order. Moreover, by multiplying the weight updates that correspond to a training pattern x with its multiplicity $mult(x)$, pattern multiplicities are easily incorporated into the learning procedure. In fact, such a technique implicitly assumes that a group of $mult(x)$ inputs is presented in succession to the network and can be considered a hybrid algorithm, as all patterns within the group cause a batch weight update (in the batch BP version, identical patterns contribute the same amount of weight change) while different groups affect learning in an on-line fashion.

This technique can also be used to balance classes in the training set when they are not equally represented. To speed up learning, one can induce larger weight changes for patterns of poorly represented categories than those of well represented categories. If N_i signifies the number of patterns in class i and $N_{max} = max\{N_i\}$, then N_{max}/N_i can be used as the amplification factor of weight changes induced by patterns from class i. For the unbalanced training set of our simulations, an increase of speed by approximately 4 times was achieved using the above technique. As is the case with the batch and on-line BP versions, selection of network parameters (e.g., adaptation gain) is problem dependent and must be performed after experimentation. However, from experience, the adaptation gain for hybrid algorithms can be selected to be the same as that for the on-line BP version without destroying stochastic convergence.

3.2   Classification Using the LVQ Algorithm

The second neural classifier used in this work is a single-layered network trained with the LVQ algorithm [16], [23]. Although the LVQ algorithm is reputably fast, further improvement on training time can be achieved by following the training set reduction procedure suggested in Section 2.2. To incorporate the multiplicities of the input patterns when updating the reference vectors, the adaptation gain $\alpha(t)$ at time t must be changed to $\alpha'(t) = [1 - (1 - \alpha(t))^{mult(x)}]$. It can easily be verified that this modification: a) is equivalent to assuming a repetitive presentation of pattern x $mult(x)$ times, considering a constant adaptation gain throughout these repetitions, and b) does not significantly affect the convergence properties of the algorithm (groups are randomly presented and the adaptation gain follows a staircase decaying function with flat portions corresponding to patterns of the same group).

3.3   The Pal-Majumder Fuzzy Classifier

The fuzzy classifier used in this work is similar to the one proposed by Pal and Majumder [13], [14] for vowel and speaker recognition and will be denoted as the PM classifier. To make it easier for readers who are unfamiliar with this, a short presentation of this algorithm follows.

Let, $x^p = (x_1^p, x_2^p,..., x_n^p)$ denote the p-th original pattern ($p = 1, ..., P$) of the training set, with x_i^p being its i-th feature element ($i = 1, ..., n$). The patterns x^p are transformed into the *fuzzy patterns* $y^p = (y_1^p, y_2^p,..., y_n^p)$ by assigning a membership function $\mu_i(\cdot)$ to each of the features x_i^p (this is called the *fuzzification* process):

$$y_i^p = \mu_i(x_i^p) = (1 + |(\bar{x}_i - x_i^p)/E|^F)^{-1} \quad \forall i = 1, ..., n, \ \forall p = 1, ..., P \quad (1)$$

where $\bar{x}_i = (1/P) \sum_{p=1}^P x_i^p$ is the mean value of the i-th feature in the training set and E, F are constants determining the shape of the membership functions (spread and steepness of the symmetric membership function).

A new test pattern, $x = (x_1, x_2, ..., x_n)$ presented to the fuzzy classifier, is first transformed to a fuzzy pattern $y = (y_1, y_2, ..., y_n)$ using equation (1) and in the sequel it is compared to each class through *fuzzy similarity scores*. Let $C_1, C_2, ..., C_C$ signify the sets of training pattern indices p that correspond to each of the C classes and let $N_1, N_2, ..., N_C$ be their respective cardinalities. The similarity scores are found by first computing the *similarity vectors* $s_c(y) = (s_{c1}, s_{c2}, ..., s_{cn}), c = 1, 2, ..., C$, between pattern y and each of the C classes, where

$$s_{ci} = (1/N_c) \sum_p s_{ci}^p \quad \forall p \in C_c \text{ and } c = 1, 2, ..., C \quad (2)$$

and

$$s_{ci}^p = (1 + W_i |1 - y_i / y_i^p|)^{-2z}. \quad (3)$$

The positive constants indicate the relative sensitivity of the classification process to the i-th feature (the lower this value, the higher the sensitivity), and z is a positive integer.

Having computed $s_1(y)$, $s_2(y)$, ..., $s_C(y)$ we then classify y to class c if $|s_c(y)| < |s_j(y)|$ $\forall j \neq c$ (this is the *defuzzification* process) with $|s_c(y)| = \sum_{i=1}^{n} s_{ci}$ for $c = 1, 2, ..., C$.

It is evident that, by classifying the SOFM prototypes instead of the original pixels ($M \times N$ spectral signatures), this algorithm can efficiently be used in multispectral satellite image classification with the proposed methodology. The reduced training sets result in a further increase in speed of the classification process by P'/P where $P' = \sum_{p=1}^{P} mult(p)$ and P are the sizes of the original and reduced training sets respectively. The only modifications needed on the original algorithm, to incorporate the group multiplicities, are in the computation of the feature means,

$$\bar{x}_i = (1/P') \sum_{p=1}^{P} mult(p)\, x_i^p, \quad \forall i = 1, 2, ..., n, \tag{4}$$

in the cardinalities $N_c = \sum_{p \in C_c} mult(p)$ and in equation (2):

$$s_{ci} = (1/N_c) \sum_p mult(p)\, s_{ci}^p \quad \forall p \in C_c \ \text{ and } \ c = 1, 2, ..., C. \tag{5}$$

3.4 Classification Using the k-NN Algorithm

The final supervised classifier used in this work is one of the simplest and most popular statistical classification methods, namely, the k-nearest neighbors (k-NN) algorithm [15]. According to the k-NN algorithm, a new input pattern x is assigned to the class voted by the majority of its k nearest (in euclidean distance sense) training patterns $x^p, p = 1, 2, ..., P$.

As was the case with neural and fuzzy classifiers, instead of classifying the original image on a pixel-by-pixel basis, the reduced training set is used to classify the catalogue of SOFM prototypes to a corresponding catalogue of labels followed by the fast indexed classification of Section 2.4. The necessary modifications to incorporate the group multiplicities of the reduced training set in the k-NN algorithm are the following: if p_i signifies the index of the i-th nearest neighbor to x then we need only find the $k' \leq k$ nearest neighbors such that

$$\sum_{i=1}^{k'} mult(p_i) \geq k \tag{6}$$

and k' has the minimum value that satisfies equation (6). Since $k' \leq k$, a further increase of the speedup factor is expected.

4   Multimodular Classification

The design of several supervised classification models with independent decisions allows for further improvement of final classification results through the use of multimodular decision making architectures. In a sense, the process of combining the power of different classifiers to obtain optimal classification results simulates the common practice followed by some patients who visit several "independent" doctors in order to obtain uncorrelated diagnoses and then follow the treatment suggested by the majority of them. By analogy, classification results obtained by a single classifier may be absolutely dependent on the particular design and properties of the classifier. Such dependence may have serious effects on final performance, especially when there is significant overlap of the categories and the optimum (in the Bayesian sense) boundaries are non-linear. Taking into consideration the computational complexity of the overall multimodular system, the increase in speed achieved by the proposed methodology in the design of each individual classifier can prove quite beneficial.

The multimodular architecture considered in this work utilizes the simple voting schemes suggested by Battiti and Colla [24]. In their experiments on optical character recognition, multiple neural network classifiers were used and independence of their individual decisions was guaranteed by using different: a) input features, b) number of hidden units, c) initial random weights, and/or d) network models. Each classifier (module) was then allocated a vote and the final decision was made by following a relative or absolute majority rule. Their results with different combinations of modules show an overall superiority of the multimodular system in terms of classification accuracy, with respect to individual classifiers.

In this work, we extend the above to multimodular classification systems that incorporate not only neural but also fuzzy and statistical classifiers. Independence of individual decisions is guaranteed by using different classification models. Absolute majority rules are then applied for a *primal* classification. Depending on the majority rule, input patterns may be rejected from classification (*don't know* cases resulting from classifier disagreement) and performances can be plotted in the *accuracy/ rejection plane*, whereby an increase in the rejection rate should result in an increase of classification accuracy provided that rejected patterns are close to class boundaries. This added flexibility offers design options that can be exploited when desired performance levels must be obtained even at the expense of an increased rejection rate.

To resolve the problem of *don't know* pixels in the primal classification result we may exploit the spatial property of the image data. To this end, a *don't know* pixel may be given the label of the majority of its local neighbors found in a window centered around the *don't know* pixel. Such a technique can be viewed as spatial *noise filtering*. This cleans the final image and homogenizes its classification regions.

5   Land-Cover Classification

In this section we apply the proposed methodology for supervised and unsupervised classification of a multispectral Landsat TM 512×512 image over Lesvos island in Greece, with a spatial resolution of 30m, into the following four land-cover categories: a) *forest*, b) *sea*, c) *agricultural* and d) *bare rock-inhabited areas*. The satellite data consisted of the first 3 bands (256 grey levels each) that correspond to the red, green and blue regions of the visible spectrum. The original image is shown in Figure 7a.

After delineation of small polygonal regions from each land-cover category by an expert, two sets consisting of 6011 and 3324 samples (3-D spectral signatures) were selected for training the supervised models and testing their classification performance, respectively. In order to assess the generalization capabilities of the supervised algorithms during training, the first of these sets was further randomly split to generate a training set of 4209 samples and a validation set consisting of 1802 samples. The number of patterns in the selected four categories of each of the three sets are shown in Table 1. For the unsupervised models, the same sets of labeled pixels can be used to assess their performance. In Section 5.3, we compare the performance of the Fuzzy Isodata automatic classifier with its supervised counter-parts.

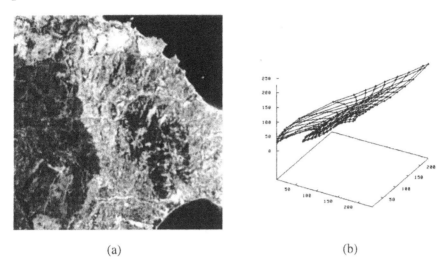

(a)   (b)

Figure 7. (a) Original multispectral satellite image used in the simulations, and (b) the self-organized map of 16×16 neurons.

Table 1. Number of patterns in each of the four categories for the three data sets.

Set	Total	Forest	Sea	Agric.	Rock
Training	4209	1375	1054	1434	346
Validation	1802	589	451	614	148
Test	3324	1098	637	944	645

All programs were run on a SUN ULTRA II Enterprise workstation (64MB, 167MHz). A map of 16×16 neurons was trained with 10^5 random presentations of the 3-D spectral signatures in 23.12 sec. The adaptation gain $\alpha(t)$ of Kohonen's SOFM algorithm was selected to tend to zero according to $\alpha(t) = \alpha_0 /(1 + K_\alpha \cdot t)$ with an initial gain $\alpha_0 = 0.3$ and a rate of decay $K_\alpha = 0.002$. The popular rectangular neighborhood function was considered, with a neighborhood size $d(t)$ shrinking with time according to $d(t) = d_{min} + d_0/(1 + K_d \cdot t)$ where $d_{min} = 1$, $d_0 = 7$ and $K_d = 0.0025$. The self-organized map produced with the above settings is depicted in Figure 7b.

Following SOFM training, the storage of asymptotic weights into the catalogue of SOFM prototypes (256 prototypes × 3 floats/prototype × 4 bytes/float $\times$ 8 bits/byte = 3×2^{13} bits) and the index table construction (512^2 indices $\times$ 8 bits/index = 2^{21} bits) required 54.30 sec.

The SOFM prototypes and index table can also be used for representing the original satellite image (3 bands $\times$ 512^2 greys/band $\times$ 8 bits/grey = 3×2^{21} bits) in a compressed form. The compression ratio achieved in this case is $3 \times 2^{21} / (3 \times 2^{13} + 2^{21}) = 2.96$ while higher ratios can be obtained for more data bands and/or smaller maps, although caution should be exercised with the latter since small maps may lead to large quantization distortions. The quantized image produced by following the indices and rounding the elements of the prototype vectors to their nearest integers is visually almost indistinguishable from Figure 7a.

5.1   Supervised Classification

The next step is to compress the training and validation sets as explained in Section 2.2. The new sets have 284 and 252 patterns respectively. This is very close to the number of SOFM prototypes. The typical strategy followed in the design of supervised classifiers is to stop training when performance on the validation set is maximized. This approach is used to avoid the well known problem of overtraining [20] and may require experimentation that involves several trials with different parameter values. In the results shown below, a slightly different strategy was used. Instead of maximizing the performance of the validation set alone, we maximized a linear combination of the training and validation performances with coefficients of the linear combination, the 0.3 and 0.7, respectively. The reason for this modification was to assure not only overtraining avoidance but also a good model performance on the training set. Parameter tuning through repetitive experimentation and assessment of training and validation performance at regular intervals adds to "real" training time and provides an additional reason for the importance of the proposed methodology.

Table 2 shows the results obtained with the BP, LVQ, PM and k-NN classifiers in terms of final performance on the training, validation and test sets of Table 1, while Table 3 shows the respective increases in training speeds achieved by the proposed methodology. A 3-10-4 feedforward network was trained with the BP algorithm using an adaptation gain of 0.5 (2.0) and momentum parameter of 0.7 (0.8), both following a staircase decay by a factor of 0.7 every 500 epochs, for the classical (proposed) methodology. Training and validation set performances were assessed every 20 epochs for both methodologies. The training times shown in Table 3 do not take into account the parameter tuning phase and correspond to 1500 epochs (6313500 presentations) for the classical methodology and 220 epochs (62480 presentations) for the proposed methodology. The increase in training speed for the BP classifier, achieved with the proposed methodology, was more than 500 with no significant change in classification accuracy.

Table 2. Performance of the four supervised classifiers for the classical and proposed methodologies (F-*forest*, S-*sea*, A-*agricultural*, R-*rock*).

Model	Category	Classical Method			Proposed Method		
		Training	Valid.	Test	Training	Valid.	Test
BP	F	98.25%	98.64%	94.72%	96.51%	97.62%	91.80%
	S	100.00%	100.00%	100.00%	99.91%	100.00%	98.27%
	A	96.44%	94.79%	93.22%	96.58%	95.11%	94.92%
	R	89.60%	91.22%	95.19%	93.64%	92.57%	97.36%
	Total	**97.36%**	**97.06%**	**95.40%**	**97.15%**	**96.95%**	**95.01%**
LVQ	F	97.16%	98.47%	93.90%	97.53%	98.64%	94.26%
	S	100.00%	100.00%	100.00%	99.91%	99.78%	98.12%
	A	97.07%	95.77%	94.70%	96.37%	95.60%	94.70%
	R	90.46%	91.22%	96.12%	89.31%	89.86%	94.26%
	Total	**97.29%**	**97.34%**	**95.73%**	**97.05%**	**97.17%**	**95.13%**
PM	F	96.58%	97.45%	93.99%	92.80%	92.02%	86.25%
	S	97.82%	96.45%	84.77%	99.62%	99.33%	89.48%
	A	94.42%	92.83%	92.06%	88.84%	87.95%	85.06%
	R	95.09%	94.59%	98.45%	94.22%	93.92%	97.52%
	Total	**96.03%**	**95.39%**	**92.54%**	**93.28%**	**92.62%**	**88.72%**
k-NN	F	97.75%	97.96%	92.35%	97.82%	98.13%	92.53%
	S	100.00%	100.00%	100.00%	99.91%	99.78%	98.12%
	A	97.84%	94.79%	92.58%	97.42%	95.60%	93.64%
	R	91.62%	92.57%	96.12%	87.57%	87.16%	93.49%
	Total	**97.84%**	**96.95%**	**94.61%**	**97.36%**	**96.78%**	**94.10%**

Training of the LVQ classifier was performed with 24 reference vectors (6 reference vectors per category) and a linearly decaying adaptation gain (initial gain equal to 0.3, decay slope = $-0.3/5000$) for both methodologies. Performances on the training and validation sets were assessed every 100 iterations. From Tables 2 and 3, we can infer that with no significant change in classifier performance, the increase in training speed achieved was about 4.5. The entries in these tables show that the LVQ algorithm is one of the most appropriate for satellite image classification due to its high speed and good generalization capabilities as long as the dimensionality of the data is relatively small (e.g., 3-D in this application).

Table 3. Increase in training speed for the four classifiers.

Classifier	Classical Method		Proposed Method		Speedup
	Number of Presentations	Time (sec)	Number of Presentations	Time (sec)	
BP	6313500	2847.47	62480	5.31	536.25
LVQ	1000	0.58	2200	0.13	4.51
PM	-	75.25	-	0.55	136.82
k-NN	-	14.57	-	0.26	56.04

Unlike the stochastic training nature of neural classifiers, the fuzzy PM and k-NN classifiers use static (non-adaptive) training. However, optimum selection of the PM and k-NN design parameters can only be achieved through repetitive classification of the training and validation sets. As with neural classifiers, optimum model selection corresponded to maximization of the combined training and validation performance index by assessing it for various sets of design parameters. In this way, the parameters selected for the PM model were $E = 0.1$, $F = 9.0$, $z = 9$ and $W_i = 0.8$ ($i = 1, 2, 3$) for the original data sets and $E = 0.1$, $F = 12.0$, $z = 12$ and $W_i = 0.8$ ($i = 1, 2, 3$) for the reduced data sets. For the k-NN classifier, $k = 5$ was found to be the optimal value of k for both the original and reduced data sets. Table 2 shows that the non-adaptive nature of these algorithms results in a worse generalization than their neural counterparts for either methodology. The increase in speed due to reduced training and validation sets was about 136 and 56 times for the PM and k-NN classifiers respectively. The training times shown in Table 3 correspond to the time needed to assess training and validation performance for a given parameter set of the PM algorithm and for $k = 5$ for the k-NN algorithm.

Finally, Table 4 shows the increase in speed achieved in classifying the original 512×512 satellite image. The times reported for the classical method correspond to pixel-by-pixel classification using those classifiers that have been trained with the original data sets. The times reported for the proposed method include the times for classifying the SOFM prototypes (20ms, 20ms, 370ms and 160ms for the BP, LVQ, PM and k-NN classifiers respectively) and the time for the indexed classification (120ms for all classifiers). The increase in speed achieved for the BP and LVQ algorithms was about 72 and 24 times respectively, while that for the PM and k-NN algorithms was about 7112 and 2926 respectively, a very impressive result. The final classification results using the original data sets are shown in Figure 8 and those for the reduced data sets are shown in Figure 9.

Table 4. Increase in classification speed of the four algorithms.

Classifier	Classical Method Time (sec)	Proposed Method Time (sec)	Speedup
BP	10.11	0.14	72.21
LVQ	3.45	0.14	24.64
PM	3485.00	0.49	7112.25
k-NN	819.38	0.28	2926.35

5.2   Multimodular Classification

The design of multimodular classification systems in such demanding applications is often prohibited by the time it takes to train the individual classifiers and then classify the data with each one in turn. For the four classifiers used in this work, the total training time (not counting the necessary repetitive trials) is 2937.87 sec while the total classification time is 4317.94 sec.

Using the proposed methodology, the total training and classification times are 6.25 sec and 1.05 sec, respectively, thus encouraging the use of such multimodular classification systems. In this work, the combined decision is based on an absolute majority voting scheme. In particular, we require at least three out of the four classifiers to be in agreement in order to accept the decision. Pixels for which there is not enough agreement (i.e., no three individual classifiers agree on a common class label) are labeled temporarily as *"don't knows"*. *Don't know* pixels are then resolved locally at the next stage by giving them the label of the majority of their labeled spatial neighbors found in a 3×3 window centered around them.

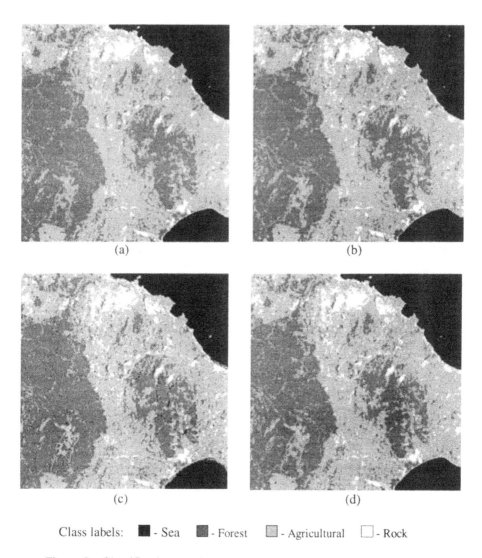

(a)   (b)

(c)   (d)

Class labels: ■ - Sea   ■ - Forest   ■ - Agricultural   □ - Rock

Figure 8. Classification results using the original data sets for the following algorithms: (a) BP, (b) LVQ, (c) PM, and (d) k-NN.

Figures 10a and 10b show the classification results obtained using a multimodular classifier on the original and reduced data sets, respectively. Qualitative evaluation and comparison with Figures 8 and 9 shows superior classification quality due to more homogeneous regions.

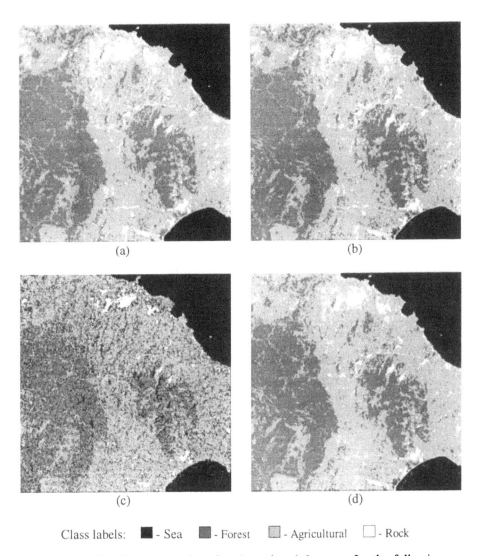

Class labels: ■ - Sea   ■ - Forest   ▨ - Agricultural   □ - Rock

Figure 9. Classification results using the reduced data sets for the following algorithms: (a) BP, (b) LVQ, (c) PM, and (d) k-NN.

The time needed to combine the decisions of the four classifiers (only for the SOFM prototypes), indexed classification, and resolving 9309 *don't know* pixels through local information was 0.15 sec for the proposed methodology. The corresponding time for the classical methodology was 0.42 sec (5790 *don't know* pixels).

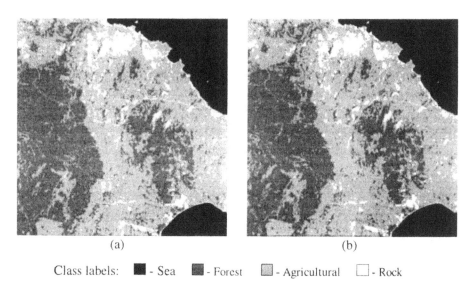

(a)   (b)

Class labels: ■ - Sea   ■ - Forest   ▨ - Agricultural   □ - Rock

Figure 10.   Multimodular classification results using: (a) the classical methodology, and (b) the proposed methodology.

5.3   Unsupervised Classification

The experiments performed in this section mainly consider two of the most popular clustering algorithms, namely the Isodata [15], [25] and Fuzzy Isodata [26] algorithms. However, in order to stress the efficiency of the proposed methodology some practical remarks on the hierarchical min-max statistical algorithm [15] have also been included.

Classification results for 8 categories (subclusters) using the proposed methodology for the Isodata and Fuzzy Isodata algorithms are shown in Figures 11a and 11b, respectively. Convergence of the Isodata algorithm was achieved in 34.71 msec (13 iterations) while Fuzzy Isodata was terminated in 0.967 sec, at 100 (preselected maximum number) iterations. The additional indexed classification time, common to all algorithms, was 0.12 sec.

Figure 11c shows the classification result obtained in 11.90 sec with the hierarchical min-max algorithm. The computational complexity prohibits direct use of this algorithm on the original data (this is a disadvantage when compared with the proposed methodology). On the other hand, direct application of the Isodata and Fuzzy Isodata algorithms to the original data is possible (see Figures 12a and 12b) at a cost of about 1024 times ($512^2/256$) longer clustering time per iteration. In fact, clustering in 8 categories required 172.63 sec for Isodata (63 iterations) and 542.31 sec for Fuzzy Isodata (50 iterations), while the time for classification was 1.02 sec for both algorithms.

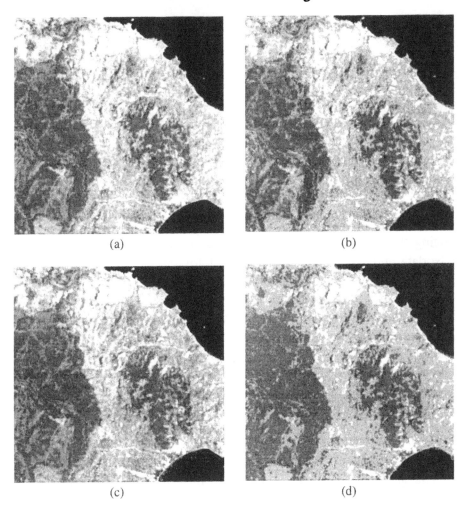

(a)   (b)

(c)   (d)

Figure 11.   Classification results in 8 categories with the proposed methodology using: (a) Isodata, (b) Fuzzy Isodata, and (c) hierarchical min-max algorithm. The classification result in 4 categories using Fuzzy Isodata with the proposed methodology is shown in (d).

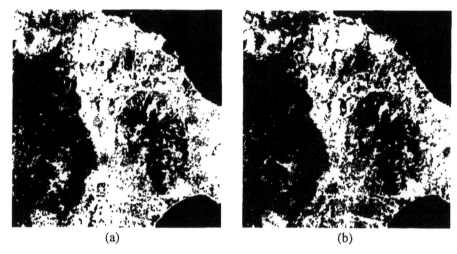

(a)   (b)

Figure 12. Unsupervised classification in 8 categories with the classical methodology using: (a) Isodata, and (b) Fuzzy Isodata.

From the above, the increase in clustering speed per iteration, achieved by using the proposed methodology, is about $2.74/(2.67 \times 10^{-3}) = 1026$ for Isodata and $10.84/(9.67 \times 10^{-3}) = 1121$ for Fuzzy Isodata (see Table 5). On the other hand, comparisons in terms of classification speed show an increase in speed, due to indexing techniques, of $1.02/0.12 = 8.5$ for both algorithms. At this point, it is important to note that if SOM training and index table construction (requiring 77.42 sec) are not off-line computations, the increase in speed in the first user trial will be smaller. However, for any additional classification trials (with different numbers of clusters) performed by the user for optimizing results, the increase in speed will be as stated above.

Table 5. Computational times and clustering gain (per iteration) for a map of 16×16 neurons. The symbol ∞ means extremely large clustering time.

Clustering Algorithm	Classical Method	Proposed Method	Speedup
Isodata	2.74 sec	2.67 msec	1026
Fuzzy Isodata	10.84 sec	9.67 msec	1121
Hierarchical	∞	11.90 sec	∞

Finally, as far as classification performance is concerned, qualitative evaluation of Figures 11 and 12 through photointerpretation shows very satisfactory results. Quantitative evaluation of the results is also possible through the labeled sets used in the supervised case. For example, Tables 6-8 show classification performances (in the form of confusion matrices) on training, validation and test data sets using the proposed methodology with the Fuzzy Isodata algorithm. Such a tabular display of the results is useful as it conveys information about the percentage of correct or incorrect data classifications per category (rows of the confusion matrix) and about class overlapping in each cluster (columns of the confusion matrix). Confusion matrices assist the user in finding the optimum number of clusters and/or merging clusters to larger ones so as to satisfy the needs of a particular application.

Table 6. Confusion matrix for the training data set.

Category	Total	SC1	SC2	SC3	SC4	SC5	SC6	SC7	SC8
F	1375	4	0	0	0	1046	0	291	34
S	1054	1043	0	0	0	0	0	11	0
A	1434	0	463	255	30	1	460	37	188
R	346	0	1	13	326	0	6	0	0

Table 7. Confusion matrix for the validation data set.

Category	Total	SC1	SC2	SC3	SC4	SC5	SC6	SC7	SC8
F	589	0	0	0	0	458	0	124	7
S	451	445	0	0	0	0	0	6	0
A	614	0	217	108	16	1	178	21	73
R	148	0	1	5	139	0	3	0	0

Table 8. Confusion matrix for the test data set.

Category	Total	SC1	SC2	SC3	SC4	SC5	SC6	SC7	SC8
F	1098	1	7	0	0	728	1	308	53
S	637	556	0	0	0	1	0	80	0
A	944	0	296	151	28	1	228	39	201
R	645	0	0	16	629	0	0	0	0

The first column of Tables 6-8 refers to the category, the second contains the number of data samples per category and the last 8 columns show the classification results for the 8 subclusters. For example, 255 out of 1434 training samples of category A and 13 out of 346 training samples of category R were classified in subcluster SC3 (see Table 6).

Next, the 8 subclusters are merged to form 4 new clusters (C1, C2, C3, C4) being in 1-1 correspondence with the 4 categories (F, S, A, R) as follows: all initial subclusters with the majority of their data in category F (i.e., columns SC5 and SC7 of Tables 6-8) are merged to form cluster C1, those with data majority in S (i.e., SC1) form cluster C2, and so on. The result of merging and the new confusion matrices are shown in Tables 9-11.

Table 9. Cluster merging for the training set.

Category	Total	C1 = F	C2 = S	C3 = A	C4 = R
F	1375	1337	4	34	0
S	1054	11	1043	0	0
A	1434	38	0	1366	30
R	346	0	0	20	326
Total	4209	1386	1047	1420	356
Performance	96.75%	97.24%	98.96%	95.26%	94.22%

Table 10. Cluster merging for the validation set.

Category	Total	C1 = F	C2 = S	C3 = A	C4 = R
F	589	582	0	7	0
S	451	6	445	0	0
A	614	22	0	576	16
R	148	0	0	9	139
Total	1802	610	445	592	155
Performance	96.67%	98.81%	98.67%	93.81%	93.92%

Table 11. Cluster merging for the test set.

Category	Total	C1 = F	C2 = S	C3 = A	C4 = R
F	1098	1036	1	61	0
S	637	81	556	0	0
A	944	40	0	876	28
R	645	0	0	16	629
Total	3324	1157	557	953	657
Performance	93.17%	94.35%	87.28%	92.80%	97.52%

The last row of Tables 9-11 shows the total classification accuracy as well as the individual accuracies in the four categories for the respective data sets. The individual accuracies are computed as 100% times the ratio of the correctly classified pixels over the total number of pixels in each category. For example, since 1366 out of the 1434 pixels of category A are correctly classified, the percentage in the last row will be 95.26%. The overall accuracy is found by adding the diagonal elements and then dividing by the total pixels in the data set. A direct comparison of the results shown in Tables 9-11 with those of Table 2 shows that unsupervised algorithms may compete in terms of performance with their supervised counterparts. However, a direct clustering to 4 categories (thus, avoiding the merging steps) would result in representing forest and sea pixels in the same category, since their spectral signatures differ less (in euclidean distance) than pixels within the agricultural category. The result would be a map with two agricultural subcategories, one category for the rock and one category for sea and forest.

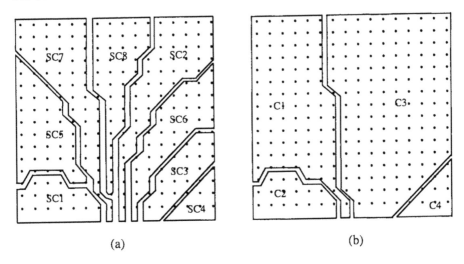

(a)   (b)

Figure 13. (a) Initial subclusters, and (b) final clusters, for a 16×16 map.

Figure 13a shows the 8 subclusters on the map that correspond to those of Tables 6-8, with dots specifying the position of neurons in the lattice. Figure 13b shows the result of merging and Figure 11d the final classification into 4 categories. Cluster/sub-cluster connectedness is a result of self-organization, while their relative position on the map follows the similarity of their spectral signatures. For example, cluster C2 that corresponds to the sea category is the exclusive neighbor of cluster C1 (forest) and, therefore, most incorrect classifications will be in the forest category rather than in the other categories. This is in agreement with the entries of Tables 9-11.

6   Summary

The methodology described in this paper offers time and memory savings for supervised and unsupervised model design and classification of large volumes of multi-dimensional spatial data using self-organizing maps and indexing techniques. The catalogue of SOFM prototypes together with the index table can be used as a compressed representation of the original data.

An increase in speed in the neural network training phase as well as in selecting the design parameters of fuzzy and statistical supervised classifiers is achieved by size reduction and redundancy removal from the training (and validation) sets in such a way as to preserve most of the information contained in the original data sets. On the other hand, an increase in clustering speed for unsupervised algorithms is achieved due to the relatively small number of SOFM prototypes that represent the original data space, permitting the use of even the most computationally complex algorithms.

Finally, efficient indexed classification, leading to increased speed, is possible by first classifying the "few" SOFM prototypes (relative to the original image data) followed by fast indirect addressing through the index table. Results on land-cover classification of multispectral satellite data show significant increases in speed of training, clustering and classification for a variety of neural, fuzzy and statistical algorithms.

References

[1] Richards, J.A. (1993), *Remote Sensing Digital Image Analysis: An Introduction*, 2nd ed., Springer-Verlag, Berlin-Heidelberg.

[2] Benediktsson, J.A., Swain, P.H., and Ersoy, O.K. (1990), "Neural network approaches versus statistical methods in classification of multisource remote sensing data," *IEEE Trans. Geosci. Remote Sensing*, vol. 28, no. 4, pp. 540-552.

[3] Bischof, H., Schneider, W., and Pinz, A.J. (1992), "Multispectral classification of landsat-images using neural networks," *IEEE Trans. Geosci. Remote Sensing*, vol. 30, no. 3, pp. 482-490.

[4] Heermann, P.D. and Khazenie, N. (1992), "Classification of multispectral remote sensing data using a back propagation neural network," *IEEE Trans. Geosci. Remote Sensing*, vol. 30, no. 1, pp. 81-88.

[5] Salu, Y. and Tilton, J. (1993), "Classification of Multispectral image data by the binary diamond neural network and by nonparametric, Pixel-by-Pixel Methods," *IEEE Trans. Geosci. Remote Sensing*, vol. 31, no. 3, pp. 606-617.

[6] Charou, E., Ampazis, N., Vassilas, N., Perantonis, S., Feizidis, C., and Varoufakis, S. (1994), "Land-use classification of satellite images using artificial neural network techniques," *Proceedings of Integration, Automation and Intelligence in Photogrammetry, Remote Sensing and GIS - LIESMARS*, Wuhan, P.R.China, pp. 368-377.

[7] Cappellini, V., Chiuderi, A., and Fini, S. (1995), "Neural networks in remote sensing multisensor data processing," *Proceedings of the 14th EARSeL Symposium'94 (Sensors and Environmental Applications of Remote Sensing)*, J. Askne (ed.), Rotterdam: A.A. Balkema, Geteborg, pp. 457-462.

[8] Narendra, P.M. and Goldberg, M. (1977), "A non-parametric clustering scheme for LANDSAT," *Pattern Recognition*, vol. 9, pp. 207-215.

[9] Goldberg, M. and Shlien, S. (1978), "A clustering scheme for multispectral images," *IEEE Trans. Systems, Man, and Cybernetics*, vol. 8, no. 2, pp. 86-92.

[10] Cannon, R.L., Dave, J.V., Bezdek, J.C., and Trivedi, M.M. (1986), "Segmentation of a thematic mapper image using the fuzzy c-means clustering algorithm," *IEEE Trans. Geosci. Remote Sensing*, vol. 24, no. 3, pp. 400-408.

[11] Wong, Y.F. and Posner, E.C. (1993), "A new clustering algorithm applicable to multispectral and polarimetric SAR images," *IEEE Trans. Geosci. Remote Sensing*, vol. 31, no. 3, pp. 634-644.

[12] Baraldi, A. and Parmiggiani (1995), F., "A neural network for unsupervised categorization of multivalued input patterns: an application to satellite image clustering," *IEEE Trans. Geosci. Remote Sensing*, vol. 33, no. 2, pp. 305-316.

[13] Pal, S.K. and Majumder, D.D. (1977), "Fuzzy sets and decision making-approaches in vowel and speaker recognition," *IEEE Trans. Systems, Man and Cybernetics*, vol. 7, pp. 625-629.

[14] Pao, Y.H. (1989), *Adaptive Pattern Recognition and Neural Networks*, Addison-Wesley, Reading, Massachusetts.

[15] Duda, R.D. and Hart, P.E. (1973), *Pattern Classification and Scene Analysis*, Wiley, New York.

[16] Kohonen, T. (1989), *Self-Organization and Associative Memory*, 3rd ed., Springer, Berlin-Heidelberg-New York.

[17] Niemann. H. and Goppert, R. (1988), "An efficient branch-and-bound nearest neighbour classifier," *Pattern Recognition Letters*, vol. 7, pp. 67-72.

[18] Lehmann, C., Viredaz, M., and Blayo, F. (1993), "A generic systolic array building block for neural networks with on-chip learning," *IEEE Trans. Neural Networks*, vol. 4, no. 3, pp. 400-407.

[19] Ienne, P., Thiran, P., and Vassilas, N. (1997), "Modified self-organizing feature map algorithms for efficient digital hardware implementation," *IEEE Trans. Neural Networks*, vol. 8, no. 2, pp. 315-330.

[20] Haykin, S. (1994), *Neural Networks: A Comprehensive Foundation*, MacMillan, Englewood Cliffs, New Jersey.

[21] Rumelhart, D.E., Hinton, G.E., and Williams, R.J. (1986), "Learning representations by back propagating errors," *Nature*, vol. 323, pp. 533-536.

[22] Fogelman Soulie, F. (1991), "Neural network architectures and algorithms: a perspective," in T. Kohonen, K. Makisara, O. Simula, and J. Kangas (eds.), *Artificial Neural Networks*, pp. 605-615, Elsevier, Amsterdam, The Netherlands.

[23] Kohonen, T., Kangas, J., Laaksonen, J., and Torkkola, K. (1992), LVQ_PAK: The Learning Vector Quantization Program Package, Helsinki University of Technology, Espoo, Finland.

[24] Battiti, R. and Colla, A.M. (1994), "Democracy in neural nets: voting schemes for classification," *Neural Networks*, vol. 7, pp. 691-707.

[25] Ball, G.H. and Hall, D.J. (1967), "A clustering technique for summarizing multivariate data," *Behavioral Science*, vol. 12, pp. 153-155.

[26] Bezdeck, J.C. (1976), "A physical interpretation of fuzzy ISODATA," *IEEE Trans. Systems, Man and Cybernetics*, vol. 6, pp. 387-389.

CHAPTER 4

LEARNING FINE MOTION
IN ROBOTICS:
DESIGN AND EXPERIMENTS

C. Versino and **L.M. Gambardella**
IDSIA, Corso Elvezia 36
6900 Lugano, Switzerland
cristina@idsia.ch, luca@idsia.ch
http://www.idsia.ch

Robotics research devotes considerable attention to *path finding*. This is the problem of moving a robot from a starting position to a goal avoiding obstacles. Also, the robot path must be *short* and *smooth*. Traditionally, path finders are either *model-based* or *sensor-based*. While model-based systems address the path finding problem *globally* using a representation of the workspace, sensor-based systems consider it *locally*, and rely only on robot sensors to decide motion. Both methods have limitations, which are rather complementary. By integrating the two methods, their respective drawbacks can be mitigated. Thus, in [15] a model-based system (a planner working on an artificial potential field) and a sensor-based system (a Hierarchical Extended Kohonen Map) which cooperate to solve the path finding problem have been described. Along related lines, several authors [5], [6], [8], [12] have proposed to build automatically the sensor-based system as the result of a learning process, where a local planner plays the role of the teacher. In particular, [5], [8] employ a Self-Organizing Map (SOM) and [6] use a dynamical variant of SOM (DSOM) based on a Growing Neural Gas network [2]. In these works, the decision of using a SOM-like network seems to be justified by its *data topology-conserving* character which is supposed to favor in some way the learning of suitable $< perception, action >$ pairs. None of these works provide experimental evidence for this reasonable, but not obvious, claim.

In this chapter we describe a SOM-like neural network which learns to associate actions to perceptions under the supervision of a planning system. By reporting this experiment the following contributions are made. *First*, the utility of using a hierarchical version of SOM instead of the basic SOM is investigated. *Second*, the effect of cooperative learning due to the interaction of neighboring neurons is explicitly measured. *Third*, the beneficial side-effect which can be obtained by transferring motion knowledge from the planner to the SOM is highlighted.

1 How to Find the Path?

A *path finder* is an algorithm to guide a robot from a starting location to a goal avoiding the obstacles in the workspace (Figure 1).

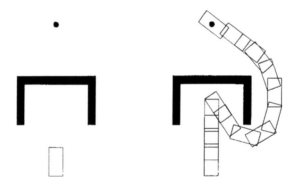

Figure 1. An instance of the path finding problem (left) and a solution to it (right). The white rectangle is the robot, the black circle is the goal, the gray shapes are the obstacles.

A good path finder generates short and smooth paths, and, for this to be possible, it requires both *high level* and *low level* motion skills. At high level, it should be able to reason on the trajectory performed by the robot as a *whole*. This is to recover from dead-ends (Figure 1) and to optimize the path length. At low level, the path finder should be able to decide on each *single step* of the robot trajectory: this is to propose actions that approach the robot to the goal while avoiding collisions. Low level strategies are also referred to as *fine motion* strategies.

Traditionally, path finders are either *model-based* or *sensor-based*. Model-based algorithms use a *model* of the workspace (a map, a camera image or other representations) to generate an obstacle-free path to the goal, while sensor-based algorithms rely only on the robot on-board *sensors* to gather information for motion. Thus, model-based systems address the path finding problem in a *global* way, while sensor-based systems consider it in a *local* way.

Advantages and drawbacks of each method are as follows.

A model-based system is able to compute short paths and to recover from dead-ends, because it works on a complete description of the workspace. However, when such a description is not available, a model-based system is of no use. Also, a model-based system requires considerable computational time to generate the actual robot path, because it needs to evaluate through simulation many alternative paths to select the best one.

A sensor-based system does not need a complete model of the workspace and this is already an advantage in itself. Also, it is computationally inexpensive as it just reacts to sensor readings. But a sensor-based system generates sub-optimal paths, and it may get trapped into dead-ends. Moreover, it is difficult to program a sensor-based system, as we have to predict every possible situation the robot will encounter, and specify a corresponding correct action.

Our research is about the integration of a model-based system and a sensor-based system so that they cooperate and solve the path finding problem in an efficient way [3], [4]. *The focus of this chapter, though, is on the automatic construction of the sensor-based system. Instead of being pre-programmed, this is shaped by a learning process where the model-based system plays the role of teacher. In this way, the programming bottleneck typical of sensor-based systems is bypassed.*

2   The Model-Based System

Traditional model-based systems solve the path finding problem by searching in the space of the robot *free configurations*, the so-called C space [11]. C is made of all robot positions and orientations that are attainable given its kinematics and the obstacles in the workspace. A robot configuration is a multi-dimensional point in C, while the obstacles are forbidden regions in the C space. The drawback of the search techniques based on C is that to determine which configurations are reachable is computationally very expensive.

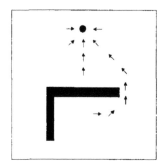 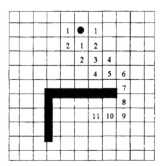

Figure 2. The artificial potential field (left), the field on a discretized workspace.

To reduce this complexity, more recent model-based approaches have suggested to search for a collision-free path by using a "direct" workspace representation of the obstacles and the robot [10]. As an example, a camera image which gives a view from above of the robot in the workspace is a 2-dimensional direct representation (Figure 1, left). Most of the techniques based on direct representations use then an *artificial potential field V* (Figure 2, left) as heuristic to guide the search in the workspace [7]: the robot just needs to follow the field V. At a given location of the workspace, V is the combination of an attractive force towards the goal and of repulsive forces exerted by the obstacles. The computation of the potential field is a one-off step, which needs to be re-executed when either the goal or the obstacles change.

The idea of following the potential field works well if the robot can be modelled as a "point". But when we consider the robot real shape and size we need to extend the potential field metaphor as described below.

First V is created on a discretized workspace (Figure 2, right) and the motion of the robot is guided by a number of *control points* c_i positioned on its shape [10] (Figure 3, left). These points can be thought of as "field sensors". Let $V(c_i)$ be the field value associated to control point c_i, whatever the robot position.

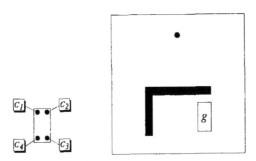

Figure 3. The control points chosen on the robot shape (left). The robot position and orientation in the workspace is its configuration (right).

At any time during motion, the robot *configuration* g is its position and orientation in the workspace (Figure 3, right). When at configuration g, the potential field value V_g of the robot is a combination of the field value of its control points (Equation 1). Roughly speaking, V_g indicates how close is the robot in attaining the goal.

$$V_g = \sum_i a_i V(c_i) \tag{1}$$

Solving the path finding problem is then equivalent to searching for a sequence of collision-free configurations that brings the robot from its initial configuration to the goal. To reduce the search space, the potential field V is used as a heuristic evaluation of the different robot configurations. In what follows, a single configuration transition is described to illustrate how the search process takes place.

Suppose g is the current robot configuration (Figure 3, right). To decide where to move the robot next, i.e., to decide its next configuration g', a set of neighboring configurations is generated by slightly varying g along the robot motion degrees of freedom. This gives rise to, say, a set of four configurations $\{g_1, g_2, g_3, g_4\}$ (Figure 4). Next, for each configuration g_i, its potential field value V_{g_i} is evaluated using Equation 1.

These values are sorted in increasing order: $\{V_{g_1}, V_{g_2}, V_{g_4}, V_{g_3}\}$. The final step is to determine which is the configuration with the smallest potential

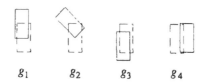

Figure 4. The neighboring configurations.

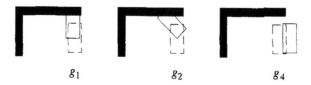

Figure 5. Collision check for robot configurations.

field value that is also collision-free. Since V_{g_1} is the smallest value, we start by simulating g_1 to check whether it produces a collision with some obstacle. Note that this verification is feasible because an image of the robot environment is available, making it possible to explore "mentally" the consequences of an action. After inspection, g_1 and g_2 are discarded (they both collide). Finally, g_4 is accepted as the next configuration for the robot: $g' = g_4$.

It may happen that all neighboring configurations of g produce collisions. This means that the robot is in a *local minimum* of V_g. Typically, a local minimum arises in a dead-end on the robot path (Figure 1). To recover from a local minimum, alternative configurations to the current one can be generated according to various rules, ranging from simply backtracking one step in the robot configuration search-tree to more sophisticated high-level motion strategies [3], [4].

We are now in a position to understand why the planner is computationally expensive. First, it calculates a great number of candidate configurations, because it cannot guess which configurations are acceptable. Second, the order of evaluation of the candidate configurations depends only on their potential values; the robot shape is not taken into account. *The planner does not learn from its own experience.*

In short, a planner is a very general method to solve path finding problems: it may be applied to robots of *any* size and shape. However, this generality is paid in terms of efficiency. When it is foreseen to work with a robot of a *given* size and shape, the flexibility provided by a planning system is less important, and the time needed to plan the robot motion is the main issue. This is why we propose to construct a *custom* sensor-based system, a system which is tailored to a robot of a given size and shape.

3 The Sensor-Based System

Suppose that a robot of a given size and shape is chosen. While using the planner to solve instances of the path finding problem, knowledge for fine motion can be collected and expressed as a set of pairs $< perception, action >$. Each pair refers to a single step of the robot trajectory, and describes a sensory perception (Figure 6) and the action selected by the planner for that perception (Figure 7). To be more specific, the perception is made of a vector o of readings of 24 obstacle proximity sensors, together with the relative goal direction g, a two-dimensional vector of unitary length. The planner action a is a triple representing a x-translation, a y-translation, and a rotation with respect to the robot's current position and orientation. Both the xy-translations and the rotation take discrete values, and can be either positive, negative or null.

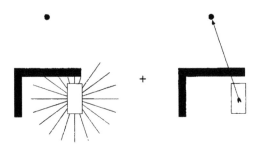

Figure 6. The robot perception: distance from the obstacles (left), relative goal direction (right).

The pairs $< perception, action >$ are the basis to build "automatically' the custom sensor-based system: for this purpose, a *supervised learning*

Figure 7. The action selected by the planner.

approach is used. A pair $< perception, action >$ is learnt incrementally by the following steps:

1. the sensor-based system gets the perception as input;

2. it generates a tentative action according to its current knowledge;

3. this action is compared to the target action selected by the planner, and the difference between the two is used to change and improve the behavior of the sensor-based system.

This sequence of steps is repeated for every example available. As a sensor-based system we use an *Artificial Neural Network* (ANN).

Before describing the ANN structure, let us observe that, in general, learning is *hard*. It cannot be achieved just by forcing through the ANN thousands of examples. To facilitate learning, we can take advantage of a good characteristic of the path finding problem, namely that *similar perceptions require similar actions*. This is not always true, but it is true most of the time. As an example, consider the two robot perceptions shown in Figure 8. The only difference between the two is a small variation in the goal position. The action "move to the right" which is suitable for the first case may also be applied to the second.

This property suggests a modular architecture for the ANN, by which its operation is logically divided into two steps (Figure 9). Given an input perception, the ANN *first* determines which is the most similar perception out of the ones experienced so far (step (A)). *Second*, it triggers the action associated to the prototypical perception selected at the first step

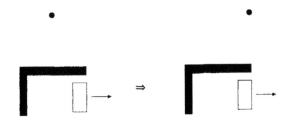

Figure 8. Similar perceptions require similar actions.

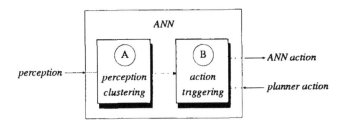

Figure 9. The two step operation of the ANN.

(step (B)). (A) is a clustering operation that aims at grouping together similar experiences into a single "prototypical perception". In this way, similar perceptions all contribute to the learning of the same (or similar) action, while different perceptions (perceptions which may require different actions) do not interfere one with the other because they are mapped to different prototypical perceptions by the clustering step.

Both step (A) and (B) require learning, and can be globally addressed by the Hierarchical Extended Kohonen Map algorithm. We start by describing step (A).

4 Perception Clustering

In (A) (Figure 9) the ANN learns the concept of similar perceptions. The task is to construct a set of prototypical perceptions out of a set of perceptions experienced by the robot during motion. This clustering task is solved by the basic *Kohonen Map* (KM) algorithm [9].

Kohonen Map. A KM is a two-layered network (Figure 10) consisting of an input layer of neurons directly and fully connected to an output layer. Typically, the output layer is organized as a two-dimensional grid G. w_r is the fan-in weight vector (reference vector) associated to the neuron placed at position r on G.

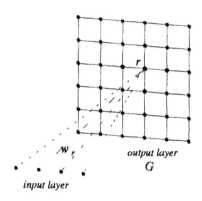

Figure 10. The KM network architecture.

The network is trained by *unsupervised* learning on a set of examples $\{x(1), \ldots, x(T)\}$. For each example x, the following sequence of steps is executed.

1. x is presented to the input layer.

2. A *competition* between the neurons takes place. Each neuron calculates the distance between its reference vector w_r and input pattern x.

$$d(x, w_r) = \|x - w_r\|^2 \tag{2}$$

   The neuron s whose weight vector is the closest to x is the *winner* of the competition.

$$s = \arg\min_r d(x, w_r) \tag{3}$$

3. s is awarded the right to learn the input pattern, i.e. to move closer to it in data space:

$$w_s^{new} = w_s^{old} + \alpha(t) \cdot \left(x - w_s^{old}\right) \tag{4}$$

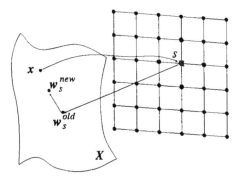

Figure 11. The KM learning step.

Figure 11 illustrates the weight change process of neuron s in the original data space.

In Equation 4, $\alpha(t)$ is the *learning rate*, a real parameter that decreases linearly with the pattern presentation number t.

$$\alpha(t) = \alpha(0) \cdot \left(1 - \frac{t}{T}\right) \qquad (5)$$

4. A special trait of the Kohonen algorithm is that the learning step is extended also to the *neighbors* of the winner s. The neighbors of s are those output elements whose distance to s, measured on the grid G, is not greater than a neighborhood parameter $n(t)$. Likewise to α, n decreases linearly with time.

$$n(t) = \left\lceil n(0) \cdot \left(1 - \frac{t}{T}\right) \right\rceil \qquad (6)$$

At the beginning of the learning process, the neighborhood parameter is large, and many neurons share the learning step. As time progresses, fewer neurons are allowed to become closer to the presented input pattern. Figure 12 shows how the neighborhood shrinks in time.

The KM can be used for *data quantization*: an input pattern x can be represented by the reference vector w_s that wins when x is presented to the KM. Also, *similar input patterns are mapped onto neighboring*

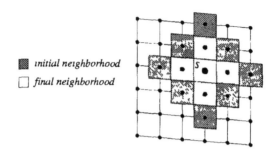

Figure 12. The neighborhood size shrinks as learning proceeds.

locations on the grid (Figure 13). This is because neurons that are close on G are also close in the original data space, a property induced by step 4 of the KM algorithm. Thus, the quantization process preserves the *topology information* contained in the original data. We will show how this topology preserving representation proactively helps the learning of fine motion.

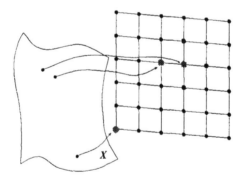

Figure 13. The KM maps similar input data to neighboring locations on the grid.

Back to perception clustering.   If a KM is trained with perceptions as inputs, then its reference vectors will represent prototypical perceptions.

To be more specific about the KM structure, remember that a perception is made of a vector o of readings of 24 obstacle proximity sensors, together with the relative goal direction g, a two-dimensional vector (Figure 6). These two vectors could be joined into a single input vector for the KM. If we do so, each neuron in the network will represent an obstacle perception and a given goal direction. However, this representation is not "economic", as it will be made clear by the following example.

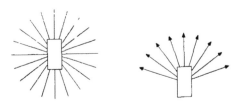

Figure 14. The robot perception while moving in the free space: obstacle perception (left), goal perceptions (right).

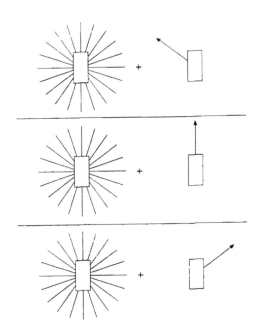

Figure 15. The reference vectors of the trained 1 × 3 KM.

Imagine the robot is moving in the free space. The only obstacle perception it can experience is that shown in Figure 14, left. In this case, the only varying input parameter is the goal direction. Suppose that the robot has experienced the goal directions shown in Figure 14, right. If we use, say, a 1 × 3 KM to cluster this input information, the training of the KM would produce the prototypical perceptions depicted in Figure 15. From this representation it is clear that the information concerning the obstacle perception is unnecessarily repeated three times.

To avoid repetitions a Hierarchical Kohonen Map (HKM) [13] may be used instead of the "flat" KM.

Hierarchical Kohonen Map. A HKM is a KM network (*super-network*) composed by a hierarchical arrangement of many subordinated KMs (*sub-networks*) (Figure 16).

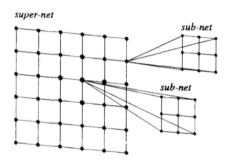

Figure 16. Architecture of the Hierarchical Kohonen Map.

A HKM is used to process input components in *sequence*.

As an example, suppose we want to process an input vector x which is the combination of two component vectors x_1 and x_2. To process x_1 and x_2 separately, we split the input presentation into two steps (Figure 17).

1. x_1 is presented as input to the super-net G. Let neuron s be the winner of the competition on x_1 in G.

2. Then, x_2 is presented as input pattern to sub-net G_s, i.e. the KM associated to the winner s in G. Now, a neuron v in G_s is selected as the winner of the competition on x_2.

The learning rule for the HKM is the same as the one presented for the simple KM. The only difference is that now we have two neighborhood parameters, one for the super-network and one for the sub-nets.

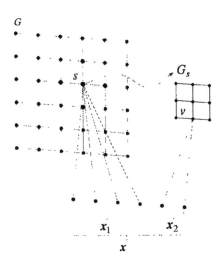

Figure 17. The HKM processes input components in sequence.

In short, in our learning case there are three reasons for preferring a hierarchical architecture to a "flat". First, it avoids unnecessary repetition of o weights for different g directions, which would be costly in terms of memory requirements. Second, it deals naturally with the economic input representation of g as a 2 dimensional vector. A flat network would need either a more distributed coding for g (as in [12]) or a weighting of g (as in [5], [6]) so that during the matching step g does not lose importance with respect to o, whose dimensionality is rather high. Third, by processing the input information in two stages, we hope to simplify the adaptation process of the SOM to the perception data distribution.

Experimental results on perception clustering.   To experiment with the HKM, we designed for the rectangle-shaped robot the workspace shown in Figure 18. In this workspace, the planner solved a number of instances of the path finding problem. Each new instance was obtained by changing the robot initial position while keeping fixed the goal position. In all, the planner generated about 500 pairs $<$ *perception, action* $>$.

The HKM super-network is a 4×6 grid of neurons, while each sub-net is an array of 10 neurons (Figure 19). For the sub-nets an array arrangement is preferred, because the data to cluster (the goal directions) are points distributed on a circle.

Figure 18. The workspace for the experiments.

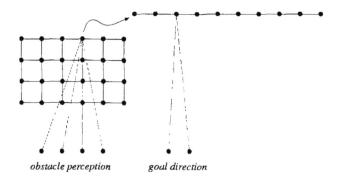

obstacle perception   *goal direction*

Figure 19. The HKM net structure.

The result of the training phase is depicted graphically in Figures 20–21.

Figure 20 shows the reference vectors of the super-network. They represent prototypical obstacle perceptions experienced by the robot during motion. For example, neuron[1] #0 is the perception of free space, neuron #5 is the perception of a wall on the right-hand side, neuron #7 is the perception of a wall behind the robot's back, and neuron #17 is the perception of a narrow corridor. Observe the *topology preserving* character of the KM: the perception similarity varies in a continuous way on the map. Therefore, similar perceptions activate neighboring locations on the grid.

---

[1]Neurons are numbered from the upper-left corner to the right.

Figure 21 shows the reference vectors of three sub-nets, namely those corresponding to neurons #0, #7, #17 in the super-network. They represent prototypical goal directions experienced by the robot for the corresponding obstacle perception. The goal direction has been represented as a white vector departing from the center of the robot shape. Again, the *topology preserving* character of the network can be appreciated: the goal direction varies on the array structure in a continuous way.

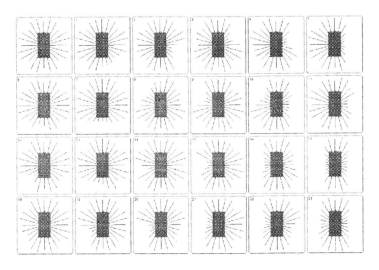

Figure 20. The obstacle perceptions learned by the super-network.

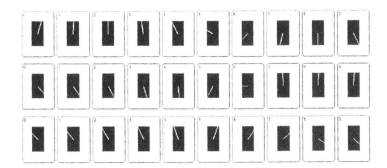

Figure 21. The goal directions learned by three sub-nets.

5   Action Triggering

In step (B) (Figure 9) the ANN learns to associate the planner actions to prototypical perceptions. This learning phase takes advantage of the ordered perception representation generated by the HKM at previous step. We introduce briefly the required extension to the basic KM algorithm (EKM) [14], which makes it possible to train the KM network by supervised learning.

Extended Kohonen Map.   From the architecture point of view, the KM network is augmented by adding to each neuron r on the competitive grid G a fan-out weight vector z_r to store the neuron output value (Figure 22).

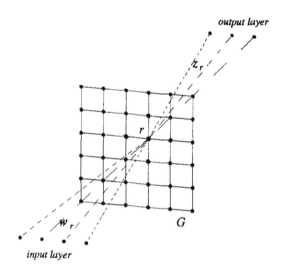

Figure 22. Extended Kohonen Map architecture.

The computation in the EKM network proceeds as follows. When an input pattern x is presented to the input layer, the neurons on G compete to respond to it. The competition involves the neurons fan-in weight vectors w_r, and consists of the computation of the distance between x and each w_r. The neuron s, whose fan-in vector w_s is the closest to x, is the winner of the competition, and its fan-out vector z_s is taken as the network output answer to x.

During the training phase, both the input pattern x and the desired output value y proposed by the teacher are learnt by the winning neuron and by its neighbors on the grid. The learning step consists of moving the fan-in weight vectors of the selected neurons closer to x, and their fan-out weight vectors closer to y (Figure 23, right).

This learning style has been described as a *competitive-cooperative* training rule [13]. It is *competitive* because the neurons compete through their fan-in weight vectors to respond to the presented input pattern. As a consequence, only that part of the network which is relevant to deal with the current input data undergoes the learning process. Moreover, neighboring locations on the grid will correspond to fan-in weight vectors that are close to each other in input data space. The rule is also *cooperative* in that the output value learnt by the winning neuron is partially associated to the fan-out weight vectors of its neighbors. If the input-output function to be learnt is a continuous mapping, then spreading the effect of the learning of an output value to the neighborhood of the winner represents a form of generalization which accelerates the course of learning [13].

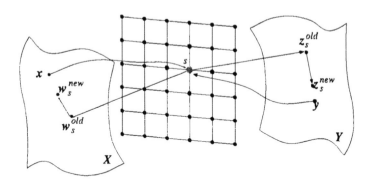

Figure 23. The learning step for the Extended Kohonen Map.

Back to action triggering.   To apply the EKM to the action learning problem, we add to every neuron in each sub-network of the HKM a fan-out weight vector that will specify the action to perform for a given obstacle perception and a given goal direction. The complete ANN architecture is a Hierarchical Extended Kohonen Map (HEKM) and is shown in Figure 24.

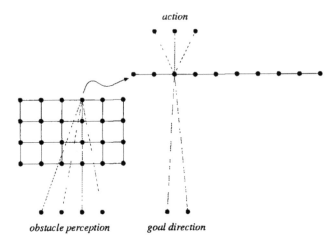

Figure 24. The HEKM network architecture.

Experimental results on action triggering. Some results for the perception clustering module have already been presented. Let us show the corresponding actions learnt by the ANN.

In Figure 25 we have represented the goal directions and the robot actions learnt by sub-network #0, the network responsible for moving the robot in the free-space. In each cell, the gray rectangle is the robot initial configuration, while the black rectangle represents the new robot configuration reached by performing the corresponding action. Similarly, Figures 26–27 show the actions learnt by the sub-networks which are responsible, respectively, to deal with the perception of a wall behind the robot's back, and that of a narrow corridor.

Again the topology preserving character of the network may be appreciated by observing how the learnt actions vary in a continuous way on each sub-network.

6   All Together

Some instances of the path finding problem are presented in Figure 28 together with the solutions found by the ANN in cooperation with the planner. In these trajectories the planner takes control only when the action proposed by the ANN would lead to a collision.

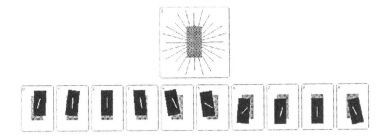

Figure 25. Obstacle perception, goal directions and actions learnt by sub-net #0.

Figure 26. Obstacle perception, goal directions and actions learnt by sub-net #7.

These paths illustrate the ANN ability to deal with several motion situations, namely avoiding a wall to the right-hand side, going around a column, entering doors, going zig-zag.

It is important to stress that, although the ANN has been trained on fine motion with respect to a *fixed* goal, the knowledge it has acquired is "general" because the goal position is not specified in absolute coordinates, but relatively to the robot. The second row of Figure 28 shows the very same ANN guiding the robot to new goal positions.

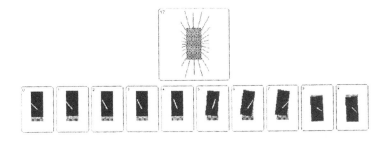

Figure 27. Obstacle perception, goal directions and actions learnt by sub-net #17.

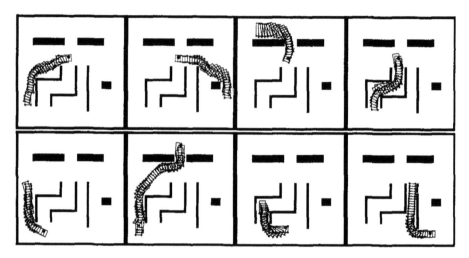

Figure 28. The robot solving the path finding problem with the fixed goal used during the training phase (first row) and with new goal positions (second row).

7   Why Use a SOM-Like Network?

We would like now to discuss the following claim: the data topology-preserving character of the HEKM favors the learning of fine motion.

This statement can be proved experimentally by performing two separate training sessions. In the first session, the neighborhood parameters (one for the super-net, one for the sub-nets) are set to 0, while in the second session they are set to values other than 0 (4 and 5, respectively). In this way, the effect of cooperation during learning can be studied.

To evaluate the two methods, an error criterion and a performance criterion are used. The error measure is the mean squared error between the network output action and the target action proposed by the planner, while the performance criterion is the percentage of optimal actions learnt by the network. By definition, the optimal actions are those proposed by the planner.

Let us comment on the plots of error and performance as a function of the number of training cycles (Figure 29). As far as the error is concerned (top plot), one can see that without cooperation (curve with black dots) a certain error level is reached quite rapidly, but afterwards, no significant improvement is observed. On the contrary, with cooperation (curve with white dots) it takes more time to reach the very same error level, but the final error is lower. This type of behavior seems to be typical for cooperating agents, as it is reported in [1]. In our experiment, a possible explanation for this could be that, when the cooperation between the neurons is active, it takes more time to find a good "compromise" to satisfy competing learning needs. However, once the compromise is met, the final result gets improved. A corresponding behavior is observed in the performance curves (bottom plot). Without cooperation a certain performance level is achieved quite rapidly (42%), but after that point no further improvement occurs. With cooperation, the same performance level is obtained later, but the final result is more satisfactory (65%).

8   Planner Vs. HEKM

We conclude by highlighting an interesting side effect which is obtained by transferring motion knowledge from the planner to the HEKM.

Our planner is a *discrete* system. By the term "discrete" we refer to the fact that, at each step of the robot trajectory, the planner generates a finite number of neighboring configurations, and chooses among these the one which approaches the goal closest while avoiding collisions. The HEKM, on the contrary, tends to produce actions which look like being *continuous*. That is because the action learnt by the network for a given perception is a kind of average action performed by the planner in similar perceptual states. To illustrate this point, we let the planner and the HEKM solve the same path finding problem as *stand-alone* systems (Figure 30). One can immediately appreciate qualitative differences in the two paths. The discrete nature of the planner is evident in the left plot: the robot motion is optimal in terms of path length, but quite abrupt. On the contrary, in the HEKM path (right plot) smoothness has been traded against optimality. This observation also accounts for the "sub-optimal" performance level reached by the HEKM (Figure 29) at the end of training.

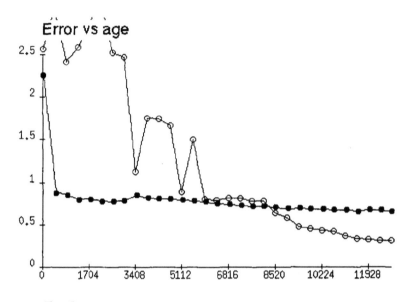

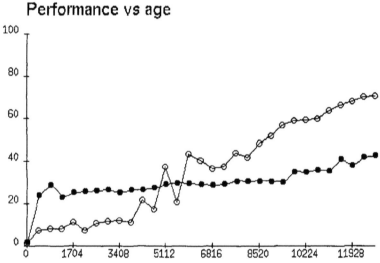

Figure 29. Error (top) and performance (bottom) without cooperation (black dots) and with cooperation (white dots).

9   Conclusions

This chapter has presented a HEKM which learns fine motion under the control of a planner.

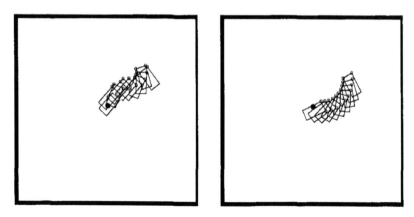

Figure 30. The planner (left plot) and the ANN (right plot) solving the same instance of path finding as stand-alone systems.

When invoked, the planner proposes an action, but to do so it needs to explore all possible robot configurations and this is expensive. This is why we capture the planner experience under the form of *<perception, action>* pairs and use these as training examples to build automatically the sensor-based HEKM. The result is that the HEKM learns to avoid collisions, acquiring an "implicit" model of the real robot through the perception-action mapping it implements. Of course the HEKM is not fault-free: when the action computed by the HEKM would lead to a collision, control is taken over by the planner, which proposes its action, and this can be treated as a new training example for the HEKM. Therefore, when the HEKM is started as a "tabula rasa", the planner is in charge of proposing the robot action most of the times. But, as the HEKM becomes more competent, control is gradually shifted to the HEKM, and the planner is invoked only in faulty cases. Overall, the integration of the planner and the HEKM improves the performance of the robot in cluttered environments because it decreases both the number of collision checks required and the number of times the planner is activated.

It is also worth noting that the integrated planner-HEKM system is able to take advantage from whatever knowledge is available about the environment. When a model of the workspace is available, we can let the planner be in charge of high-level navigation while the HEKM takes care of fine motion. But if the environment is unknown, we can still use the trained sensor-based HEKM as a stand-alone system to control the robot because the HEKM does not rely on a model of the workspace. The only constraint is to perform the training of the HEKM in a known environment so that the planner can act as a teacher. Then, as we have shown, most of the fine motion knowledge acquired in one environment can be transferred to another one.

This chapter also provides answers to a number of questions concerning the design and properties of the HEKM. *First*, we discussed the utility of using a hierarchical KM instead of the usual "flat" version. The HEKM is more economic in terms of the way memory cells are used. It avoids unnecessary weight repetitions and allows for compact input representations. Clearly, one limitation of the current architecture is the fixed number of neurons. A growing network, able to increase the number of neurons where a more detailed representation is needed, could be used instead [2], [6]. *Second*, we measured the effect of cooperative learning due to the interaction between adjacent neurons. We found that with cooperation learning is slowed down on the short run. But the benefits appear later on, resulting in a more satisfactory final performance. Our interpretation is that, at the beginning of learning, neighboring neurons work to meet a compromise to competing needs: this effort becomes rewarding on the long run. *Third*, we pointed out the complementary nature of the paths generated by the planner and by the HEKM as standalone systems. The HEKM produces sub-optimal but smooth solutions, whereas the planner seeks for optimality while sacrificing the continuity of motion. The integration of these two attitudes leads to good results.

Acknowledgments

Cristina Versino was supported by the project No. 21-36365.92 of the Fonds National de la Recherche Scientifique, Bern, Suisse.

References

[1] Clearwater, S.H., Hogg, T., and Huberman, B.A. (1992), "Cooperative problem solving," *Computation: The Micro and the Macro View*, Huberman, B.A. (Ed.), World Scientific.

[2] Fritzke, B. (1995), "A growing neural gas network learns topologies," *Advances in Neural Information Processing Systems 7*, Tesauro, G., Touretzky, D.S., and Leen, T.K. (Eds.), MIT Press, Cambridge, MA, pp. 625-632.

[3] Gambardella, L.M. and Versino, C. (1994), "Robot motion planning. Integrating planning strategies and learning methods," *Proc. AIPS94 - The Second International Conference on AI Planning Systems*, Chicago, USA, June.

[4] Gambardella, L.M. and Versino, C. (1994), "Learning high-level navigation strategies from sensor information and planner experience," *Proc. PerAc94, From Perception to Action Conference*, Lausanne, Switzerland, September 7-9, pp. 428-431.

[5] Heikkonen, J., Koikkalainen, P., and Oja, E. (1993), "Motion behavior learning by self-organization," *Proc. ICANN93, International Conference on Artificial Neural Networks*, Amsterdam, The Netherlands, September 13-16, pp. 262-267.

[6] Heikkonen, J., Millán, J. del R., and Cuesta, E. (1995), " Incremental learning from basic reflexes in an autonomous mobile robot," *Proc. EANN95, International Conference on Engineering Applications of Neural Networks*, Otaniemi, Espoo, Finland, August 21-23, pp. 119-126.

[7] Khatib, O. (1986), "Real-time obstacle avoidance for manipulators and mobile robots," *The International Journal of Robotics Research*, vol. 5, no. 1.

[8] Knobbe, A. J., Kok, J. N., and Overmars, M.H. (1995), "Robot motion planning in unknown environments using neural networks," *Proc. ICANN95, International Conference on Artificial Neural Networks*, Paris, France, October 9-13, pp. 375-380.

[9] Kohonen, T. (1984), *Self-Organization and Associative Memory*, Springer Series in Information Sciences, 8, Heidelberg.

[10] Latombe, J.-C. (1991), *Robot Motion Planning*, Kluwer Academic Publishers.

[11] Lozano-Perez, T. (1982), "Automatic planning of manipulator transfer movements," *Robot Motion*, Brady et al. (Eds.), The MIT Press.

[12] Millán, J. del R. (1995), "Reinforcement learning of goal-directed obstacle-avoiding reaction strategies in an autonomous mobile robot," *Robotics and Autonomous Systems*, vol. 15, no. 3, pp. 275-299.

[13] Ritter, H., Martinetz, T., and Schulten, K. (1992), *Neural Computation and Self-Organizing Maps. An Introduction*, Addison-Wesley Publishing Comp.

[14] Ritter, H. and Schulten, K. (1987), "Extending Kohonen's self-organizing mapping algorithm to learn ballistic movements," *Neural Computers*, Eckmiller, R. and von der Marlsburg, E. (Eds.), Springer, Heidelberg.

[15] Versino, C. and Gambardella, L.M. (1996), "Learning fine motion by using the hierarchical extended Kohonen map," *Proc. ICANN96, International Conference on Artificial Neural Networks*, Bochum, Germany, 17-19 July, pp. 221-226.

CHAPTER 5

A NEW NEURAL NETWORK FOR ADAPTIVE PATTERN RECOGNITION OF MULTICHANNEL INPUT SIGNALS

M. Fernández-Delgado[1,2], **J. Presedo**[1], **M. Lama**[1], and **S. Barro**[1]

[1]Department of Electronics and Computer Science
University of Santiago de Compostela
15706 Santiago de Compostela, Spain
[2]Department of Informatics
University of Extremadura
Cáceres, Spain
eldelga@usc.es

MART (Multichannel ART) is a neural computational architecture which is based on ART architecture and aimed at pattern recognition on multiple simultaneous information input paths. This chapter describes the characteristic aspects of MART architecture, how it operates in pattern recognition and its flexibility in adapting itself to the temporal evolution of input information fed into the network. Finally, a real application is presented demonstrating its potential for solving complex problems, above all in the field of multichannel signal processing.

1   Introduction

The neural computational model ART was introduced by Carpenter and Grossberg in the 1980s, and developed through various versions, such as ART1 [5], ART2 [4], ART-2A [8] and ART3 [6]. These networks have contributed a number of valuable properties with respect to other neural architectures, amongst which could be mentioned its on-line and self-organizing learning. On the other hand, these networks make it possible to resolve the dilemma between plasticity and stability, allowing both the updating of the classes learned and the immediate learning of new classes without distorting the already existing ones. This property enables it to be used in problems in which the number of classes is not limited a priori, or in which there is an evolution in the classes over time. These characteristics are shared by a great number of variants of the ART architecture, which are acquiring more operational and application possibilities. Thus, mentioning only a small number of the most representative examples, many networks have been proposed such as ARTMAP [9], which enables supervised learning, Fuzzy ART [7] and Fuzzy ARTMAP [10], which adapt themselves to the processing of fuzzy patterns, or HART (Hierarchical ART) [3], integrated by a series of ART1 networks which carry out cascade clustering tasks on input patterns. Nevertheless, this wide range of options lacks certain characteristics which, to our way of thinking, are especially interesting in order to tackle certain problems typical to pattern recognition such as:

- In many pattern recognition applications there are various paths or channels of information about the same event or system under analysis. This is the case, for example, with monitoring of systems by means of multiple sensors that supply complementary or alternative views of the system behavior. In these cases, the joint consideration of the information given by all of the sensors enables us to increase the reliability of the final result, making it easier to detect noise and eliminate ambiguities.

- Supervised learning does not permit the reconfiguration of a neural network during its routine application. In the case of ART networks, however, the representations of the classes to which belong the input patterns are constructed on-line, starting from a total absence of a priori information on these patterns and their associated classes. Nevertheless, this type of network does not adapt the capability of discrimination between classes of patterns during its operation, since they operate with a fixed vigilance parameter. Thus they cannot adapt their behavior to the typical characteristics of each class to be recognized nor to their possible temporal evolution.

- ART networks carry out a partitioning or clustering operation on the input space, on the basis of a vector of features that describe the input patterns, a measure of the distance between these patterns and the classes to be discerned, and a (vigilance) threshold to be applied to the distance obtained between the input pattern to be classified and previously recognized classes. Each one of these classes (associated to a cluster in the input space) includes a series of patterns. Thus, it is possible for classes to exist whose patterns are much more similar amongst themselves than with those associated to other classes. ART networks presume that the patterns of all classes have the same variability, as they use a single vigilance for all of them.

- In many pattern recognition problems the different classes have a different representativeness about the input patterns. The relevance of a class usually depends on the specific criteria of the problem tackled (appearance frequency of input patterns associated to this class, level of danger associated to this class, etc.). ART networks do not supply any mechanisms for selectively evaluating the different classes learned.

The relevance, which to our understanding these properties have, leads us to propose a new model of neural computational architecture, which we have called MART [14]-[16]. Maintaining those aspects of ART architecture that are especially relevant, we have incorporated the capability for the adaptation of those parameters that determine the operation of the network, adapting its values to the application and to the set of patterns to be processed. At the same time, MART deals jointly, although in a selective and adaptive manner, with multiple channels of information and internal representations of the classes learnt from the input data submitted to the network during its operation. In the following section we will describe the structure and operation of MART, in order to later deal with its properties for the learning of information. We will then present an illustrative example of the operation on artificially generated patterns, in order to later deal with its application in a real problem such as the recognition of morphological patterns of heart beats on multiple derivations of electrocardiographic signals. Finally we will summarize the main contributions of MART to neural computation, along with the future lines of study on which our work is focused.

2   Architecture and Functionality of MART

As we said before, MART is a neural computation architecture for the pattern recognition, which operates simultaneously over multiple channels or inputs of information. Figure 1 shows its structure, made up of I blocks, each one of which is associated to the processing of the input pattern in a different signal channel. Each one of these blocks determines the local similarity (relative to channel i) between the input pattern and the expected values of the learnt classes, through processing of the patterns showed previously to the network in that channel. The $F4$ block, located on the top of the figure, is governed by a competitive *"winner-takes-all"* mechanism, in which the only active output, $u_k=1$, is associated to the class with maximum global (over the set of channels) similarity (we will call this class "winning class"). This class propagates its expected value downwards through the single channel blocks, with the aim of determining the local differences, relative to each channel, with the input pattern. The Orientation System evaluates these differences, determining whether they are or are not sufficiently low in order to assign the input pattern to the winning class (resonance or reset respectively). Finally, the Class Manager controls the dynamic creation and suppression of classes as it becomes necessary in the classification process.

2.1   Bottom-Up Propagation in a Single-Channel Block

Figure 2 shows the structure of a single-channel block of MART. The input pattern in channel i, $E_i=(E_{i1},...,E_{iJ})$, $E_{ij} \in [0,1]$, $\forall j$, is presented in the units of layer $F1_i$, where it is propagated towards $F2_i$. The connection weight vector $z_{ik}=(z_{i1k},...,z_{iJk})$ between the units of $F1_i$ and unit $F2_{ik}$ stores the expected value of class k in channel i. The output of unit $F2_{ik}$ is determined by the expression:

$$L_{ik} = f(\mathbf{E}_i, \mathbf{z}_{ik}), \ \forall k \tag{1}$$

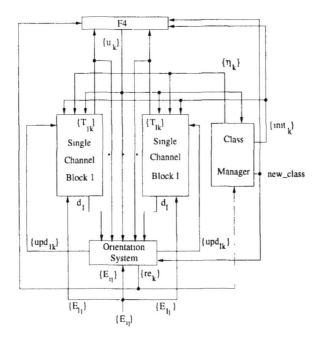

Figure 1. Block diagram of MART.

where the function $f(x,y)$ evaluates the dissimilarity between the vectors x and y, $0 \le f(x, y) \le 1$, being 0 the value associated to a total coincidence between them. The vector $L_i = (L_{i1},...,L_{iK})$ is propagated towards layer $F3_i$, whose output is:

$$T_{ik} = \eta_k (1 - L_{ik}), \quad \forall k \tag{2}$$

In this expression, η_k is an output item of the Class Manager (see Figure 4) which verifies $\eta_k = 1$ when the unit k in $F4$ is committed (i.e., associated to a learnt class), otherwise $\eta_k = 0$.

2.2 Class Selection

The input items for block $F4$ (local similarities) are integrated for the different channels generating the global similarities P_k:

$$P_k = \overline{re}_k \sum_{t=1}^{I} T_{tk} \tag{3}$$

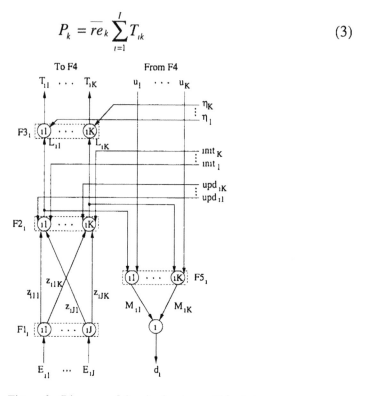

Figure 2. Diagram of the single-channel block i.

The input items re_k come from the Orientation System and, as will be seen, their value is 1 for those classes which have been temporarily reset for having shown an excessively high difference with the input pattern, for which $P_k = 0$. During the first comparison between the input pattern and the learnt classes no unit is reset, in such a manner that $P_k = \sum_{t=1}^{I} T_{tk}$, $\forall k$. The binary output items u_k of F4 are governed by the expression:

$$u = init_k \wedge \left\{ \overline{new\_class} \wedge \left[\bigwedge_{l<k} \tau(P_k - P_l) \right] \wedge \left[\bigwedge_{l>k} \Gamma(P_k - P_l) \right] \right\}, \ \forall k \tag{4}$$

where the symbols $\vee$ and $\wedge$ designate the OR and AND binary operators, respectively, and the functions $\tau(x)$ and $\Gamma(x)$ are governed by the following expressions:

$$\tau(x) = \begin{cases} 0 & x \le 0 \\ 1 & 0 < x \end{cases} \qquad\qquad \Gamma(x) = \begin{cases} 0 & x < 0 \\ 1 & 0 \le x \end{cases}$$

The *new_class* and *init*=(*init_1*,...,*init_K*) input items are binary, and come from the Class Manager: *new_class*=1 is associated to the creation of a new class, because the input pattern does not resonate with any of the learnt ones. The *init* vector is only non-zero when *new_class*=1, in which case its only non-zero component is the one associated to the unit *k'* selected by the Class Manager to be initialized. Thus, if *new_class*=1, then u_k=*init_k* (from expression (4)) and the unit *k'* is the only active unit in F4. Otherwise, *new_class*=0=*init_k*, $\forall k$, and the value of u_k will be determined by the functions τ and Γ, from expression (4), being zero except for the minimum index unit with the maximum P_k (a class which shows the greatest global similarity with the input pattern). We will denote this unit by means of the index k_1, due to it being the winner in this first cycle of comparison between the input pattern and the classes learnt: in this way, $u_{k_1}=1$, $u_k=0$, $\forall k \neq k_1$.

2.3 Top-Down Propagation in a Single-Channel Block

The vector $u=(u_1,...,u_K)$ is propagated towards the $F5_i$ layers of the *I* single-channel blocks, whose output takes the form of:

$$M_{ik} = u_k L_{ik}, \quad \forall k \qquad\qquad (5)$$

These output items are zero, except for the unit k_1, for which it is $M_{ik_1}=L_{ik_1}=f(E_i,z_{ik_1})$ (difference between the input pattern and the expected value of the active class k_1 in channel *i*). Finally, the output d_i of the single-channel block *i* is determined according to the expression:

$$d_i = \sum_{k=1}^{K} M_{ik} = L_{ik_1} = f(\mathbf{E}_i, \mathbf{z}_{ik_1}) \qquad\qquad (6)$$

which represents the difference between the input pattern and the expected value of class k_1 in that channel. This output is propagated towards the Orientation System, whose structure and functioning will be dealt with in the following section.

2.4   The Orientation System

The basic objective of this module, as shown in Figure 3, is to determine the global result of the pattern-class comparison. For this reason, the block "Channel Credits" determines the global difference d between the input pattern and a given class from the local differences d_i and the channel credits x_i, which constitute an indirect measure of the signal quality in the different channels. The global difference is calculated through the following expression:

$$d = \sum_{i=1}^{I} x_i d_i \tag{7}$$

The value obtained for the global difference d is compared with the global vigilance $\rho_{kl}{}^g$ associated to the class k_l (one of the K outputs of the block "Global Vigilances", in Figure 3). This parameter establishes the discrimination capability of the system in the comparisons with the k_l class, having an adaptive value which is limited to the range $\rho_{min} \leq \rho_{kl}{}^g \leq 1$. This comparison takes place in the block "Reset Evaluation", which determines the output items re_k of the Orientation System based on the expression:

$$re_k(t+1) = \left[re_k(t) \wedge \overline{new\_pattern(t)} \right] \vee \left[\overline{re_k(t)} \wedge u_k \wedge \Gamma(d - \rho_k^g) \right], \quad \forall k \tag{8}$$

in such a manner that if $d < \rho_{kl}{}^g$ and $re_{kl}(t)=0$ then $re_{kl}(t+1)=0$ and the state of the system, with the unit k_l active in $F4$, is maintained stable, reaching resonance with the input pattern. In this case, as shall be seen in the following section, the information associated with the resonant class k_l is updated and the resonance is maintained until the presentation of a new input pattern. On the other hand, if $d \geq \rho_{kl}{}^g$ and $re_{kl}(t)=0$, then $re_{kl}(t+1)=1$ and, from the expression (3), $P_{kl}=0$, class k_l being inhibited for the competition in $F4$ (re_{kl} remains active until the presentation of a new input pattern, the instant in which the new_pattern(t) output of block "Detection of New Pattern" is activated, taking again re_{kl} to zero). When this occurs, a new winner k_2 is determined in *F4* and the comparison cycle is repeated, re_{k2} being evaluated according to $\rho_{k2}{}^g$ and the new value of d. If the result is a new reset, the process is repeated until either a resonance with one of the learnt classes is reached, or until all are reset, in which case the Class Manager determines the creation of a new class. This class will be, if in *F4* there exist uncommitted nodes (not associated to learnt classes), the one with the minimum k' index. If there is no uncommitted node, it will select the class with minimum relevance (determined through its class credit), as we will see in the next section.

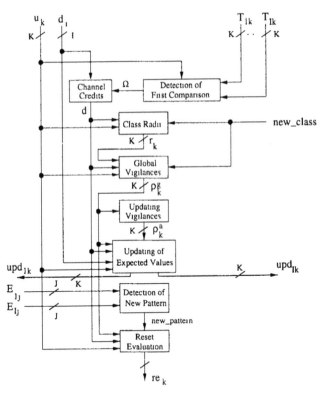

Figure 3. Structure of the Orientation System. For clarity the figure does not show the neural organization of the different blocks.

2.5   Class Manager

Figure 4 shows the structure of the Class Manager block, which controls the creation and dynamic suppression of classes throughout the processing, being activated when the input pattern belongs to an as of yet unlearned class. In this case, all the classes learned beforehand (for which $\eta_k=1$) have been reset ($re_k=1$) and the "Creation of a New Class" block activates its *new_class* output, based on the following expression:

$$new\_class = \bigwedge_{k=1}^{K} \overline{\eta_k} \vee re_k \qquad (9)$$

When *new_class=1*, the "Class Selection" block determines the class k' which is going to be created, establishing $init_{k'}=1$, $inic_k=0$, $\forall\ k\neq k'$. Other input data to this block are the credits of class μ_k ($k=1,...,K$), which evaluate the associated relevance of the different classes. As previously mentioned, these parameters allow us to perform a selective evaluation of the classes learned according to their relevance. In a later section we will describe the rule which governs the evolution of the

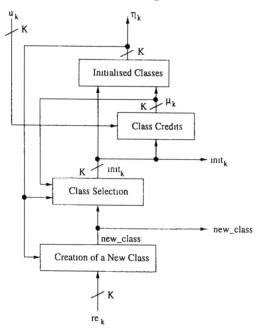

Figure 4. Structure of the Class Manager.

class credits; for the present suffice it to know that $\mu_k \in [0,1]$, $\mu_k=1$ being the credit associated with the maximum relevance. Finally, the input data η_k also take part in the creation of a new class (we should remember that $\eta_k=1$ for those units k of *F4* associated to learned classes). The output of the "Class Selection" block is determined by the following expression:

$$inic_k = new\_class \wedge \left[\bigwedge_{l<k} \tau(\varsigma_k - \varsigma_l) \right] \wedge \left[\bigwedge_{l>k} \Gamma(\varsigma_k - \varsigma_l) \right], \; \forall k \qquad (10)$$

where ς_k is defined in the following manner:

$$\varsigma_k = (1-\mu_k)\xi + \overline{\eta_k}, \; \forall k \qquad\qquad \xi = \bigwedge_{l=1}^{K} \eta_l$$

and the functions $\tau(x)$ and $\Gamma(x)$ are the ones previously defined. In this manner if $new\_class=0$, $init_k=0$, $\forall k$; i.e., a new class is created only when $new\_class=1$. In this case, if there exists some unit in $F4$ that is not associated to learned classes (for which $\eta_k=0$), then $\varsigma_k = \overline{\eta_k}$ and $init_k=1$ for this unit (if there are various, the unit selected is the one with the minimum index k). On the contrary, if all the units in $F4$ are associated to learned classes ($\eta_k=1$, $\forall k$), $\varsigma_k=(1-\mu_k)$ and the class selected is the one with the minimum class credit μ_k, i.e., the one that has the least relevance at that time.

At this point we end the description of the structure and working of MART in order to tackle aspects associated with its plasticity and adaptation according to the information extracted for the input patterns over time. This description is included in the following section.

3   Learning in MART

The MART network uses an unsupervised, *"on-line"* learning, typical of ART networks, for the determination and updating of the expected values of the different classes which are being identified during the processing. Nevertheless, it also provides other learning mechanisms, including channel credits and credits, radii and global vigilances associated to the classes learnt, as will be seen in this section.

3.1   Expected Values

The expected value z_{ik} of class k in channel i is updated with each input pattern which matches this class. Nevertheless, in order to avoid possible classification errors provoking strong distortions in this expected value, it is desirable to use a threshold ρ_k^u (updating vigilance), more restrictive than global vigilance ρ_k^g, and to update only in those channels i in which the local difference d_i is lower than the global vigilance. The following rule governs class expected value learning:

$$z_{ijk}(new) = inic_k E_{ij} + \overline{inic_k}\left[(1-\alpha_z upd_{ik})z_{ijk}(old) + \alpha_z upd_{ik}E_{ij}\right], \quad \forall i,j,k \tag{11}$$

When a new unit k is created, $init_k=1$ and $z_{ijk}(new)=E_{ij}$. In the case of resonance with class k, z_{ik} only evolves if $upd_{ik}=1$, which occurs when $d<\rho_k^u$ in those channels i where $d_i<\rho_k^g$. We can see that, if $upd_{ik}=1$, then we have, from the expression (11):

$$z_{ijk}(new) = (1-\alpha_z)z_{ijk}(old) + \alpha_z E_{ij} \quad \forall i,j,k \tag{12}$$

where the parameter $0\leq\alpha_z\leq1$ determines the speed of change in the expected value of the resonant class.

3.2   Channel Credits

Channel credits, associated to the weights inside the block "Channel Credits", in the Orientation System (see Figure 3), have an initial value of $x_i=1/I$, $\forall i$. The credit of a channel represents its weight in the global classification developed by MART over every input pattern. Channel credits update according to the result of the comparison between the pattern presented to the network and the class which has results the most similar to it. This measure of similarity does not take into account the own channel credits, and the most similar class, k_l, will be that with greatest $\sum_{i=1}^{I}T_{ikl}$, during the first comparison (in this stage the output Ω of the block "Detection of First Comparison", in the Orientation System, is activated). In effect, except for specific cases associated with the appearance of patterns which belong to classes not learnt by the network yet, the local difference in each channel with the most similar class should be reduced. In order to do this, a channel with repeatedly high local differences d_i can be considered to be associated with a higher noise content and lower signal quality, and as such, its credit should be lowered. On the contrary, if the local differences in this first comparison are reduced, this can be considered as a reliable channel with regard to the classification process, and as such, its credit should be increased. In this way, an indirect estimation of the noise-to-signal ratio is used to determine the credit or weight factor of each channel in the integration and evaluation of multi-channel information.

Obviously, the full functionality of the channel credits is reached when there is a temporal continuity in the input signals, which allows to use the information associated to previous times in order to make a prediction, in this case, about the signal quality in every signal channel. Learning is carried out on the basis of the following expression:

$$x_i(new) = \Theta\big[x_i(old) + \Delta x_i(d_i)\big], \; \forall i \tag{13}$$

The function $\Delta x_i(x)$ determines the value of the increase in x_i on the basis of the following expression:

$$\Delta x_i(x) = \begin{cases} \Delta x\left[1 - \dfrac{x}{\delta_{i1}}\right] & 0 \le x < \delta_{i1} \\[2mm] -\Delta x\left[\dfrac{x - \delta_{i1}}{\delta_{i2} - \delta_{i1}}\right] & \delta_{i1} \le x < \delta_{i2} \\[2mm] -\Delta x & \delta_{i2} \le x \le 1 \end{cases} \tag{14}$$

where Δx is a fixed amount, independent of the signal channel. In this way, the increase $\Delta x_i(x)$ is positive for reduced values of x ($0 \le x \le \delta_{i1}$) and negative for high values of x ($\delta_{i1} \le x \le 1$), reproducing in this manner the behavior outlined previously. The parameters Δx, δ_{i1} and δ_{i2}, $\forall i$, must be determined for each application, if we have a set of input patterns which are representative of those which will be presented to the network during its normal operation. Lastly, the function $\Theta_i(x)$ limits the values of x in the range $0 \le x \le 2/l$.

3.3   Class Radii

For each class learnt, MART establishes an adaptive average of the
global differences which have been obtained with those patterns
assigned to that class over the time. This average is an approximate
measurement of the radius of the cluster associated to the class in the
input space and, as will be seen, constitutes the basis for the updating of
the global vigilance for this class. Class radii correspond to the weights
inside the block "Class Radii", in the Orientation System (see Figure 3).
The learning of the radius associated to a class takes place with each
resonance between a new input pattern and that class or, for the first
time, when this class is created, and it is governed by the following
expression:

$$r_\lambda(new) = \overline{u_\lambda} r_\lambda(old) + u_\lambda \overline{\{new\_class}[(1-\alpha_r)(r_\lambda(old) + \delta(r_\lambda(old))d) + \alpha_r d]\} \quad \forall k$$
$$(15)$$

where the function $\delta(x)$ is 1 if $x=0$ and 0 in the opposite case. The
radius remains constant for non-winning classes ($u_k=0$). If
$new\_class=1$, $r_k(new)=0$ for the class k' created, since in this case
$u_{k'}=1$. On the other hand, in the case of resonance, radius is equal to the
global difference, if $r_k(old)=0$, otherwise being updated as a weighted
sum of its previous value and the global difference d (expression (16)),
weighted by a variation factor α_r. In this manner, the radius of a class is
adapted to the variability existent between the patterns belonging to it.

$$r_k(new) = (1-\alpha_r)r_k(old) + \alpha_r d, \quad \forall k \qquad (16)$$

3.4   Global Vigilances

The learning of global vigilances for each class allows the adaptation of
the system's discrimination capacity to the level of variability of the
input patterns. For this, for each resonance MART compares the global
difference d with the radius r_k of the resonant class. The increments in
the variability are translated into increases in d with respect to r_k, and it
is then advisable to increase ρ_k^g in order to avoid false negatives when
the input pattern is not assigned to class k and creates a redundant class.
On the contrary, reductions in variability reduce d against r_k, which
allows the reduction of the global vigilance in order to adapt the

discrimination capability to the new situation. Figure 5 shows an example of the time evolution in the variability of the input patterns. Initially ($t=0$) the variability is small, and the vigilance is low for that class, from the radius associated to it. Later ($t=N$), this variability grows up, and it is advisable to increase the vigilance in order to avoid that input patterns belonging to that class lead to the creation of redundant classes. Finally ($t=M$), the variability of the input patterns reduces again, which decreases the radius associated to that class and its vigilance.

Global vigilances are associated to the weights of the block "Global Vigilances", in the Orientation System (see Figure 3). The expression that controls the learning of global vigilances is the following:

$$\rho_k^g (new) = \overline{u_k} \rho_k^g (old) + u_k \left\{ \rho_{ref} \ new\_class + \right.$$
$$\left. + \theta \left[\rho_k^g (old) + sgn(d - r_k) \Delta \rho \right] \overline{new\_class} \right\} \ \forall k \qquad (17)$$

where the functions $sgn(x)$ and $\theta(x)$ are defined by the following equations:

$$sgn(n) = \begin{cases} -1 & x < 0 \\ 0 & x = 0 \\ 1 & x > 0 \end{cases} \qquad \theta(x) = \begin{cases} \rho_{min} & x < \rho_{min} \\ x & \rho_{min} \le x < 1 \\ 1 & 1 \le x \end{cases}$$

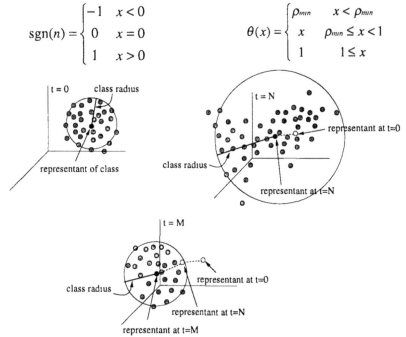

Figure 5. Illustration of the temporal evolution of the variability of the input patterns associated to a class and updating of its representative.

Parameter $\rho_k{}^g$ remains constant, except in the creation of a new class k', in which case $\rho_k{}^g$ has the initial value ρ_{ref}, and when resonance is achieved with class k''. In this case, $\rho_{k''}{}^g$ increases its value in $\Delta\rho$ if $d>r_{k''}$ and decreases in the same amount if $d<r_{k''}$, maintaining a lower limit ρ_{min} which is associated to the maximum discrimination capability that a class may possess. In this way, MART is able to selectively evaluate the different classes learnt by means of an individualized and adaptive consideration in its discrimination capability.

3.5   Other Characteristics

As commented in Section 1, another path for the selective evaluation of the classes learnt is associated with "class credits", which corresponds with the connection weights inside the block "Class Credits", in the Class Manager (see Figure 4). These credits allow the evaluation of the relevance of each class throughout the operation of the network. The credit μ_k associated to class k has an initial value $\mu_k=1$ at the moment of its creation, increasing in a constant factor Δ_p with each input pattern assigned to it, and decreasing in a constant factor Δ_n in the opposite case, always within the range $0\leq\mu_k\leq1$.

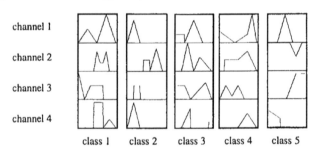

Figure 6. Classes used in the generation of patterns.

On the other hand, MART dynamically manages the classes learnt, creating a class when faced with the appearance of a pattern not belonging to any of them, or deleting a class when its credit is set to zero. For this reason, the output items η_k from the block "Initialized Classes", in the Class Manager (see Figure 4), allow discrimination between those committed units in $F4$ ($\eta_k=1$) and those which are not assigned to a class yet. For each unit k, $\eta_k=1$ when it is committed, maintaining this value while its credit μ_k is not zero. When $\mu_k=0$, this class is "forgotten" and unit k remains uncommitted until it is re-committed to a new class, thus enabling the dynamic creation/ elimination of classes as and when necessary.

4   Analysis of the Behavior of Certain Adaptive Parameters

In order to illustrate the operation of MART we now give an application example on a set of 2,000 artificially generated patterns with $I=4$ signal channels, a pattern length of $J=125$ values in each channel, and input data in the range $0 \leq E_{ij} \leq 1$. These patterns were generated from 5 basic morphologies, each one of these being labeled as belonging to a class identified with the morphology from which each pattern derived. The distance function $f(x, y)$ used is the city-block distance between the input pattern x and the representative of class y. Figures 6 and 7 show the original classes and some of the patterns generated from them, respectively.

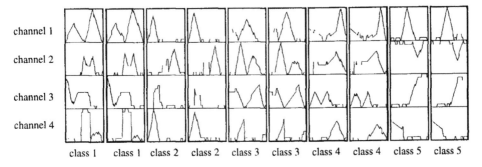

Figure 7. Examples of patterns.

Figure 8 shows an example operation, although in this case, for reasons of clarity, it is represented on 2 signal channels, the credit x_2 being notably lower than x_1. The lower left-hand section shows the input pattern in $F1_1$ and $F1_2$. This pattern propagates to $F2_1$ and $F2_2$: the competitive process in $F4$ selects class 1 as the one that is most similar to the input pattern, its representative (expected value) being shown in both channels with the upper left-hand section. At the same time, the right-hand section shows the area resulting from subtracting the input pattern and the expected value of the winning class in both channels, together with the channel credits and global difference and vigilance. The area of difference (local difference) is lower in channel 1 and somewhat higher in channel 2. Nevertheless, the reduced credit value x_2 attenuates the contribution of channel 2 to the global difference d, the value of which is lower than the global vigilance associated to class 1, and brings about a situation of resonance with the input pattern.

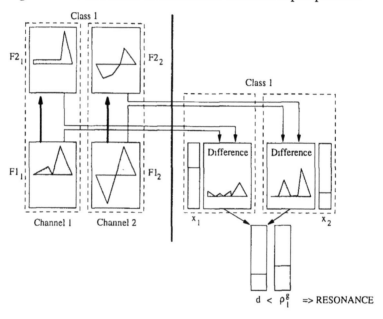

Figure 8. Example of resonance between an input pattern and its most similar class.

Figure 9 shows a second example in which the input pattern undergoes a reset with its most similar class (class 2). The latter wins the competitive process in *F4*, but the descending propagation generates a greater difference in channel 2, the credit of which is higher than the one associated with channel 1, leading to a situation of reset (first comparison). In the second comparison class 3 is the winning class in *F4*. The descending propagation provokes a higher difference in channel 1, but lower in channel 2, reaching resonance on the basis of the weighting relative to the respective channel credits.

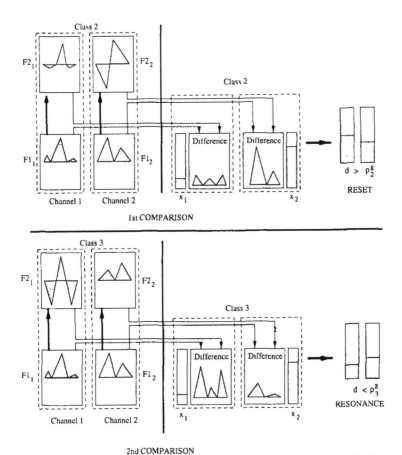

Figure 9. Example of reset between an input pattern and its most similar class, and the subsequent resonance after the comparison of the pattern with another class.

With the aim of illustrating the working of the channel credits x_i in the MART operation we distorted the input pattern, although only in channel 1, by means of the substitution of a part of the original input pattern E_{ij} for randomly generated values. Figure 10 demonstrates how the appearance of noise leads to significant rises in the local differences in channel 1 in the first comparison, differences which are thus useful in order to estimate the high noise/signal ratio associated to this channel. The figure also demonstrates the reduction in the channel credit x_1, which is produced in those intervals where noise is added to the signal (elevations in d_1) suitably reflecting the drop in reliability in channel 1 with regard to the final classification process of the input patterns. We should also emphasize the high level of stability in those credits associated with the remaining channels, in which the signal quality does not undergo any noticeable difference throughout the process.

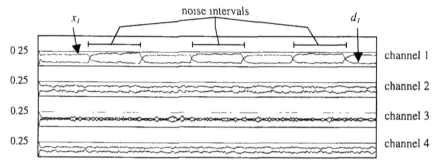

Figure 10. Evolution of the channel credits x_i in response to the addition of noise in channel 1.

Another interesting aspect of the operation of MART is associated with the evolution of the global vigilances associated to the different classes. In order to illustrate this point we added noise to the patterns belonging to class 1 at certain instances of the processing. Figure 11 shows the evolution of the vigilances $\rho_1{}^g$ and $\rho_2{}^g$; it can be seen how addition of noise leads to a noticeable increase in the global difference associated with those patterns that resonate with class 1; the one associated with class 2 remains approximately constant. These increases lead to a rise in the radius associated with class 1 (r_1) and, thus, in the vigilance $\rho_1{}^g$ of this class, whilst $\rho_2{}^g$ undergoes no significant alterations. In turn, the disappearance of the added noise brings about an immediate decrease in r_1 and $\rho_1{}^g$. Consequently, this figure shows the capability of MART's vigilance to adapt itself to the properties of the input data, more specifically to the variability that these patterns demonstrate over time.

5   A Real Application Example

In the previous section we analyzed the operation of MART on a set of artificial patterns, with the aim of highlighting some of the most interesting characteristics of learning in MART. We now show, albeit briefly, the application of MART in a real problem.

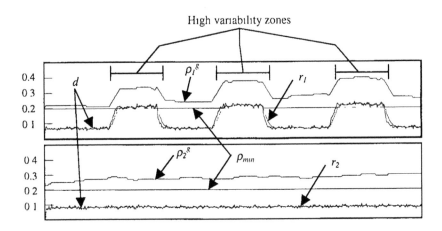

Figure 11.   Evolution of the global differences, radii and global vigilances associated to classes 1 and 2.

ART networks have indeed been employed in a wide variety of problems. Amongst these we could mention the extraction of rules from massive data for weather forecasting [21], recognition of aerial images [20] or written characters [23], identification of patterns in turbulent fluids [17], creation of semantic associations between terms on a text database [24], detection of patterns in satellite images [28], recognition and retrieval of aircraft parts in databases [12], etc. There also exist applications aimed at the automatic monitoring of signals in chemical plants [29] and nuclear power stations [22].

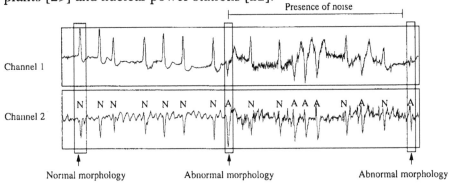

Figure 12.   ECG interval on 2 derivations in which beats with normal (N) and aberrant (A) morphologies are indicated. A signal interval with noise present is also indicated.

Finally, one of the principal application environments of these networks is biomedicine, especially cardiovascular medicine, where we find examples aimed at the extraction of information from ECG signals ([26],[27],[18]) or the prediction of the risk of myocardial infarct on the basis of clinical and electrocardiographic data [13]. It is in this setting that the problem we deal with in this section is taken, the recognition of morphological patterns in heart beats, more specifically from ventricular activation complexes or QRS complexes, on multiple ECG derivations ([1],[15]). Figure 12 shows a signal interval over 2 derivations, in which different beats can be seen. Those beats associated to normal cardiac activity (N) have similar morphologies (normal morphology). On the other hand, cardiac complication generally become apparent at the electrocardiographic level, associated to beats with aberrant morphologies, such as the ones labeled "A" in the figure. The presence of different points of origin of the heart beats generally means an unfavorable prognosis for the patient, due to which it is of great importance to recognize this situation in real time. Different origins normally give rise to different morphologies in the electrical manifestation of the beat associated to them, due to which it is essential to detect all the beat morphologies that are produced over time and classify each new beat detected in accordance with them. In Figure 12, a signal interval with significant noise can also be seen; this distorts the morphology of beats and makes its correct morphological characterization difficult.

Although this problem belongs to the already classic field of pattern recognition, it has a series of typical properties which make it especially complex and, on the other hand, suitable for the use of MART in its resolution:

- The morphological characterization of beats on various ECG derivations is very important, given that it significantly increases the information available in the single-channel case. This, then, is a problem for which the use of a multichannel pattern recognition system such as MART is suitable.

- Heartbeat morphologies vary substantially between different patients and according to the derivation of the ECG under consideration, and even for the same patient over time. Thus it is impossible to construct a sufficiently representative training set, which renders supervised approaches inadequate for this problem.

- Any new morphology is important and should be learned immediately, since it may reflect the appearance of complications in the mechanism of generation and propagation of beats. Furthermore, any morphology may change substantially throughout processing. All this requires the learning of new classes as well as the adaptation of the ones that have already been learned, i.e., to resolve the dilemma between stability and plasticity. This need precludes solutions involving off-line learning, making ART architecture-based networks prime candidates for this problem.

- The determination of a reduced set of representative characteristics of the heart-beat morphology may turn out to be extremely complex, which would make operation directly onto the ECG signal itself prudent.

- The signal may be contaminated by noise from different origins, which may significantly alter the morphology of the beats. Thus the pattern recognition system must be able to adapt its discrimination capability according to the noise level or, alternatively, to the morphological variability shown by the input patterns at each instant. As they use a fixed vigilance during the processing, ART networks do not allow such an adaptation, as opposed to MART, which does so by means of its capability to adapt its vigilance parameters.

- Classes formed by morphological patterns that are very similar amongst themselves may co-exist with others made up of patterns with a high level of variability in their morphology. As a consequence, it is advisable to adapt the discrimination capability of the system to the typical characteristics of each class. MART demonstrates this property, as it uses a different vigilance for each class, with an evolution that depends on the patterns assigned to it.

- The appearance of artifacts that imitate true beats leads to the creation of spurious morphological classes, which should be rejected. With this aim, MART's ability to dynamically suppress classes is interesting, because it favors the elimination of those classes with the lowest degree of representativeness (class credit) amongst the input patterns.

We now go on to describe how this problem has been resolved using MART. Our data set consists of a group of 20 electrocardiographic registers from the MIT-BIH Arrhythmia Database [19]. This database was chosen due to it being widely known, and due to the high number of beat morphologies that it contains. Furthermore, a morphological labeling of its beats is supplied, which allows a rigorous validation of the solution applied to the problem. The input to MART in each channel (the MIT-BIH database has $I=2$ signal channels) is $J=128$ ECG samples corresponding to an interval of 256 msec., which included the whole of the QRS complex associated with each of the beats analyzed. The maximum number of classes to be learned was set at $K=15$. Amplitude scaling was performed on the input patterns so that its maximum value was I and its minimum value 0. The distance measure used $f(x,y)$ is the city-block distance between the input pattern in each channel and the expected value of each class, as is shown in the following expression:

$$f(\mathbf{E}_i, \mathbf{z}_{ik}) = \frac{1}{J} \sum_{j=1}^{J} \left| E_{ij} - \frac{z_{ijk} - \min_{l=1, \ldots J}\{z_{ilk}\}}{\max_{l=1, \ldots J}\{z_{ilk}\} - \min_{l=1, \ldots J}\{z_{ilk}\}} \right| \tag{18}$$

This distance, traditionally used in electrocardiographic pattern recognition [11], was chosen due to its intuitive character (area between the vectors to be compared), its relative ease of calculation and the existence of works that prove its superiority with regard to other distance measurements on the ECG signal [25]. It should be borne in mind that the expected value of the class k in the channel i, $z_{ik}=(z_{i1k}, \ldots, z_{iJk})$, is not scaled to $[0,1]$, as occurs with the input, given the learning rule associated to it (expression (12)), which is the reason why it is necessary to include this scaling into the distance calculation. Figure 13 shows how the morphologies of the expected values of the different morphological classes are codified in the weights z_{ijk} associated to each one of them, being used in order to determine the distance with the input data.

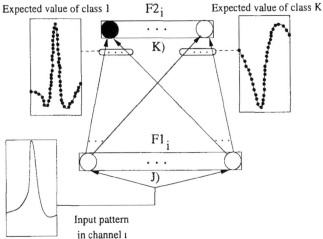

Figure 13. Values of the connection weights z_{ijk}, where the expected value of class k in channel i is stored. The shaded unit in $F2_i$ is the one that is associated to the class whose expected value has the greatest morphological similarity with the input pattern in channel i.

It is important to point out that, given that MART dynamically creates and suppresses classes, the classes created by MART do not generally coincide with those determined by the database. In this sense, the appearance of noise usually leads to the creation of redundant classes (which occurs when beats belonging to the same morphological class in the database divide into two or more classes in the MART network) although the use of adaptive channel vigilances and credits help to reduce this phenomenon.

Figure 14 shows the evolution of the channel credits and local differences, together with the number of classes created since the beginning of the processing of the register 105 of the MIT-BIH database. It can be seen that the principal increment in the number of classes is produced in instants (1) and (2), where noise appears in both channels. Nevertheless, the first of the increments, which occurs when x_1 is high, is notably greater than the second one, where x_1 has already been reduced in response to the reduction in quality in channel 1.

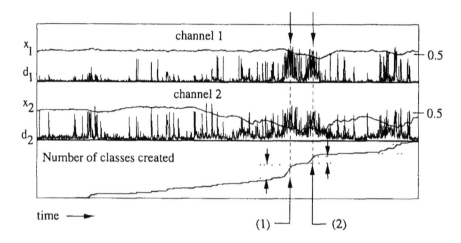

Figure 14. Temporal evolution of channel credits, local differences and the number of classes created by MART on register 105 of MIT-BIH Arrhythmia Database.

Figure 15 shows the number of classes created by MART, with and without channel credits, throughout the processing of register 200 of the MIT-BIH database. Both figures make it evident that the use of channel credits to a great degree prevents the proliferation of classes by lowering the contribution to the final result of less reliable or noisy channels. In this case, the low and high ranges for the local differences are determined by the parameters $\delta_{11}=\delta_{21}=0.05$ and $\delta_{12}=\delta_{22}=0.20$, where $\Delta x=0.0025$.

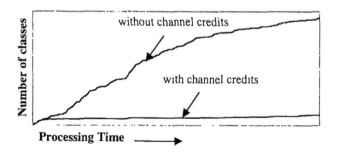

Figure 15. Temporal evolution of the number of classes created by MART on register 200, with and without channel credits.

With regard to the evolution of the class vigilances, the values used were $\rho_{min}=\rho_{ref}=0.15$, with $\Delta\rho=0.002$. More specifically, the adaptation capability of the vigilances was another reducing factor in the number of classes created by MART, as can be seen in Figure 16, by relaxing the discrimination capability in the presence of noise, thus preventing the creation of redundant classes.

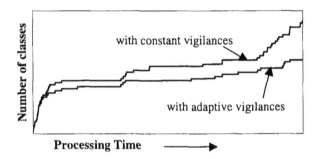

Figure 16. Temporal evolution of the number of classes created by MART during the processing of register 200, using constant and adaptive vigilances.

Finally we will briefly comment on the results obtained in the application of MART to this problem ([1],[15]). The most revealing piece of data is the lower percentage of characterization errors (1.1%), which illustrates MART's high capacity for morphological discrimination. This parameter only measures errors due to the assignation of beats to classes to which in reality they do not belong, considering any beat assigned to a redundant, but morphologically similar class to be correct. Another illustrative parameter is the redundant class creation ratio: the number of classes created by MART is on average 5.7 times higher than the real number. Although this ratio may appear to be high, it is in fact reasonable, bearing in mind, on one hand, the discrimination capability that is necessary in order to distinguish between classes, and, on the other hand, the noise level, and the high level of variability associated to some of them. However, the percentage of error due to the assignation of beats to redundant classes is 23%, in spite of the number of classes created by MART being almost six times higher than the original one. This means that the majority of the beats (over 76%) are assigned to the class to which they belong, it becoming evident that there is a noticeable capacity for limiting the population of redundant classes. Lastly, we would point out that the use of MART for the morphological pattern recognition on multichannel ECG is operative at the present inside a monitoring system of patients in a ICCU developed by our research group ([2],[15]).

6   Discussion

We do not wish to finish off this chapter without having emphasized some of the principal characteristics that MART contributes in context of neural networks and, more specifically, in that of networks based on the ART model. Amongst these, there are its express orientation towards the recognition of multichannel patterns and its ability for the continuous adaptation to the characteristics of the input pattern. In this sense, besides the obvious capacity for learning and updating of the classes to be discerned, MART offers greater operational flexibility than other neural networks aimed at pattern recognition, allowing the dynamic suppression of classes, selective evaluation of the different signal channels according to the signal quality at each moment in time and adaptation to the variability of the patterns associated to the different classes.

MART's adaptation possibilities reach the height of their functionality in the operation on patterns associated to signals that evolve continually over time. It is here where, for example, the channel credits are associated with an indirect measurement of the signal/noise ratio. These characteristics of MART, which make it unique in the field of neural computation, endow this architecture with an important practical scope. This has become evident when tackling a real and complex problem, such as that of the real time morphological characterization of beats on multichannel ECG.

Lastly, our current objectives include developing new MART-based applications, principally in the field of processing signals of a physiological origin, an area in which our group has wide experience, and to continue improving the performance of MART with regard to its capabilities for the recognition of multichannel patterns and on-line learning guided by network input patterns.

Acknowledgments

Authors wish to acknowledge the support from Xunta de Galicia, under project XUGA20608B97, and from the Spanish Ministry of Education and Culture (CICYT), under project 1FD97-0183.

References

[1] Barro, S., Fernández-Delgado, M., Vila, J.A., Regueiro, C.V. and Sánchez, E., (1998), "An adaptive neural network for the classification of electrocardiographic patterns on multichannel ECG," *IEEE Eng. in Medicine and Biology Magazine*, vol. 17, pp. 139-150.

[2] Barro, S., Presedo, J., Castro, D., Fernández-Delgado, M., Fraga, S., Lama, M. and Vila, J., (1999), "Intelligent tele-monitoring of critical patients," *IEEE Engineering in Medicine and Biology Magazine*, vol. 18, no. 4, pp. 80-88.

[3] Bartfai, G., (1996), "An ART-based modular architecture for learning hierarchical clusterings," *Neurocomputing*, vol. 13, pp. 31-45.

[4] Carpenter, G.A. and Grossberg, S., (1987), "ART2: Self-organizing of stable category recognition codes for analog input patterns," *Applied Optics*, vol. 16, no. 23, pp. 4919-4930.

[5] Carpenter, G.A. and Grossberg, S., (1988), "The ART of adaptive pattern recognition by a self-organizing neural network," *Computer*, vol. 21, pp. 77-87.

[6] Carpenter, G.A. and Grossberg, S., (1990), "ART3: Hierarchical Search Using Chemical Transmitters in Self-Organizing Pattern Recognition Architectures," *Neural Networks*, vol. 3, pp. 129-152.

[7] Carpenter, G.A. and Grossberg, S., (1991), "Fuzzy ART: Fast stable learning and categorization of analog patterns by an adaptive resonance system," *Neural Networks*, vol. 4, pp. 759-771.

[8] Carpenter, G.A., Grossberg, S. and Rosen, D.B., (1991), "ART-2A: An Adaptive Resonance Algorithm for Rapid Category Learning and Recognition," *Neural Networks*, vol. 4, pp. 493-504.

[9] Carpenter, G.A., Grossberg, S. and Reynolds, J.H., (1992), "ARTMAP: supervised real-time learning and classification of nonstationary data by a self-organizing neural network," *Neural Networks*, vol. 4, pp. 565-588.

[10] Carpenter, G.A., Grossberg, S., Markuzon, N., Reynolds, J.H. and Rosen, D.B., (1992), "Fuzzy ARTMAP: A neural network architecture for incremental supervised learning of analog multidimensional maps," *IEEE Transactions on Neural Networks*, vol. 3, no. 5, pp. 698-712.

[11] Cashman, P., (1978), "A pattern-recognition program for the continuous ECG processing in accelerated time," *Computers and Biomedical Research*, no. 11, pp. 311-323.

[12] Caudell, T.P., Smith, S.D.G., Escobedo, R. and Anderson, M., (1994), "NIRS: large scale ART-1 neural architectures for engineering design retrieval," *Neural Networks*, vol. 7, no. 9, pp. 1339-1350.

[13] Downs, J., Harrison, R.F., Kennedy, R.L. and Cross, S.S., (1996), "Application of the fuzzy ARTMAP neural network model to medical pattern classification tasks," *Artificial Intelligence in Medicine*, vol. 8, pp. 428-493.

[14] Fernández-Delgado, M. and Barro, S., (1998), "MART: A multichannel ART-based neural network," *IEEE Transactions on Neural Networks*, vol. 9, pp. 139-150.

[15] Fernández-Delgado, M., (1999), *MART: una nueva arquitectura de computación neuronal y su aplicación al diseño de un sistema inteligente de monitorización de pacientes*, Ph.D. Thesis, Dept. of

Electronics and Computer Science, University of Santiago de Compostela. In Spanish.

[16] Fernández-Delgado, M., Presedo, J. and Barro, S., (1999), "Multichannel pattern recognition neural network," *Proceedings of the International Work-Conference in Artificial and Natural Neural Networks (IWANN-99)*, Alicante (Spain), pp. 719-729.

[17] Ferre-Gine, J., Rallo, R., Arenas, A. and Giralt, F., (1996), "Identification of coherent structures in turbulent shear flows with a fuzzy ARTMAP neural network," *International Journal of Neural Systems*, vol. 7, no. 5, pp. 559-568.

[18] Ham, F.M. and Han, S., (1996), "Classification of cardiac arrhythmias using ARTMAP," *IEEE Transactions on Neural Networks*, vol. 43, no. 4, pp. 425-430.

[19] Harvard-MIT division of Health Sciences and Technology, "The MIT-BIH arrhythmia database CD-ROM," (1992), 2nd ed.

[20] Healy, M.J., Caudell, T.P. and Smith, D.G.S., (1993), "A neural architecture for pattern sequence verification through inferencing," *IEEE Transactions on Neural Networks*, vol. 4, no. 1, pp. 9-20.

[21] Healy, M.J. and Caudell, T.P., (1997), "Acquiring rule sets as a product of learning in a logical neural architecture," *IEEE Transactions on Neural Networks*, vol. 8, no. 3, pp. 461-473.

[22] Keyvan, S., Durg, A. and Nagaraj, J., (1997), "Application of artificial neural networks for the development of a signal monitoring system," *Expert Systems*, vol. 14, no. 2, pp. 69-79.

[23] Kim, H.J., Jung, J.W. and Kim, S.K., (1996), "On-line Chinese character recognition using ART-based stroke classification," *Pattern Recognition Letters*, vol. 17, pp. 1311-1322.

[24] Muñoz, A., (1996), "Creating term associations using a hierarchical ART architecture," *Lecture Notes on Computer Science*, Springer-Verlag, vol. 1112, pp. 171-177.

[25] Rappaport, S.H., Gillick, L., Moody, G.B. and Mark, R.G., (1982), "QRS morphology classification: quantitative evaluation of different strategies," *Computers in Cardiology*, pp. 33-38.

[26] Suzuki, Y. and Ono, K., (1992), "Personal computer system for ECG ST-segment recognition based on neural networks," *Medical & Biological Engineering & Computing*, vol. 30, pp. 2-8.

[27] Suzuki, Y., (1995), "Self-organizing QRS-wave recognition in ECG using neural networks," *IEEE Transactions on Neural Networks*, vol. 6, no. 6, pp. 1469-1477.

[28] Waldemark, J., (1997), "An automated procedure for cluster analysis of multivariate satellite data," *International Journal of Neural Systems*, vol. 8, no. 1, pp. 3-15.

[29] Whiteley, J.R., Davis, J.F., Mehrotra, A., and Ahalt, S.C., (1996), "Observations and problems applying ART2 for dynamic sensor pattern interpretation," *IEEE Transactions on Systems, Man and Cybernetics*, Part A: Systems and humans, vol. 26, no. 4, pp. 237-423.

CHAPTER 6

LATERAL PRIMING ADAPTIVE RESONANCE THEORY (LAPART)-2: INNOVATION IN ART

T.P. Caudell
Department of Electrical and Computer Engineering
and
The Albuquerque High Performance Computing Center
University of New Mexico
Albuquerque, N.M. 87131
U.S.A.
tpc@eece.unm.edu

M.J. Healy
Phantom Works
The Boeing Company
PO Box 3707 Mail Stop 7L-66
Seattle, Washington 98124-2207
U.S.A.
Michael.J.Healy@boeing.com

In this chapter, we present the results of a study of a new version of the LAPART adaptive inferencing neural network [1], [2]. We will review the theoretical properties of this architecture, called LAPART-2, showing it to converge in at most two passes through a fixed training set of inputs during learning, and showing that it does not suffer from template proliferation. Next, we will show how real-valued inputs to ART and LAPART class architectures are coded into special binary structures using a preprocessing architecture called Stacknet. Finally, we will present the results of a numerical study that gives insight into the generalization properties of the combined Stacknet/LAPART-2 system. This study shows that this architecture not only learns quickly, but maintains excellent generalization even for difficult problems.

1   Introduction

A Holy Grail of neural networks is *fast learning with good generalization*. In many neural architectures, these two trade off against each other, making it difficult to achieve them simultaneously. In this chapter, we present a version of the LAPART adaptive inferencing neural network architecture [1]-[3] that has excellent learning and generalization properties. LAPART architectures are constituted from two or more ART architectures bilaterally connected with adaptive connections. The centerpiece of the chapter is the theorem that under certain broad conditions, LAPART-2 converges in at most two passes or epochs through a fixed set of binary training inputs, where an epoch is the single-time application of a complete list of input patterns to a neural network for learning. In [4], Georgiopoulos, Heileman and Huang proved the upper bound $n-1$ on the number of epochs required for convergence for the similar ARTMAP architecture, where n is the size of the binary pattern input space; they also proved that the bound can decrease with increasing vigilance parameter values, ρ. ARTMAP performs a function similar to LAPART; both require binary-valued input patterns although, as we will show, they can process real-valued input patterns in a manner equivalent to that of Fuzzy ARTMAP through the use of stack interval pre-processing networks [6]. The ARTMAP result can be thought of as an n pass, or finite-pass, convergence result. In these terms, LAPART-2 is then a 2-pass, or fixed-pass, convergence result. To our knowledge this is the first fixed-pass convergence result of its kind.

LAPART-2 is a byproduct of theoretical and empirical investigations into the learning properties of the previous version, LAPART-1. Based upon formal modeling of the semantics of neural networks [7], LAPART-1 was developed specifically to learn logical inference relationships, or rules, between classes of objects from an application. Both the classes and the rules are formed in the synaptic memory of the network according to neural design principles embodied in ART-1 [8], but with a logical design principle from the formal semantic analysis: adaptive neural network connections implement logical implication [7], [9]. The logic, however, changes to adapt to the data. The underlying reason for this is that many inferences made by a neural network prove to be unsound when tested on new data, so the logic must be corrected.

We regard this principle as a point of departure not only for theories of learning with neural networks, but for learning in general.

The overarching question facing us is the following: Can a neural network – ART-1 or LAPART, for example – adapt its logic so that, from some point in time forward, its inferences are valid provided that it is presented with data sufficiently similar to that which it has previously experienced? A positive answer has been provided for ART-1 with a learning parameter set within a range of values commonly used [10]: ART-1 converges on a fixed training set of binary input patterns in a number of presentation epochs that can be calculated from information about the data. The required information is the number N of *different sizes* of input patterns, where the size of a binary-valued pattern is the number of 1-valued components it possesses. If all input patterns have the same size, then only a single epoch is required. This is an N-pass convergence result for unsupervised learning in terms of the stratification of the input space into size classes. This result is especially interesting in that it specializes to one-pass convergence for fixed input pattern size.

The inferences made by an ART-1 network are simply its self-organized classification decisions: The nodes in the F2 (classification) layer of the network compete in a winner-take-all fashion, and a binary *template* pattern comprising adaptive connections from the winner to the F1 (comparison) layer is compared with the input pattern to determine if the input belongs in the corresponding class of patterns. Thus, the classes have two representations: F2 nodes and templates. An ART-1 network has converged on a fixed training set of patterns when all inputs have a *direct access* template in the system – one that causes immediate classification. The point here is that the logic of a trained ART-1 system is valid not only for the training patterns, but for any future input patterns that have direct access templates in the ART-1 memory. However, conditions for other ART-type architectures, in particular LAPART-1, are difficult to derive because of the phenomena that can occur in the more complex situations of inference learning. Assumptions must be made, either upon the architecture in the form of added design constraints or upon the data in the form of "domain constraints." The former is the basis for the design change to create LAPART-2, and the latter is the basis for the hypothesis that makes our two-pass convergence theorem possible.

The LAPART-1 architecture has been described in [1], [2], [3]. It couples two ART-1 networks, designated subnetworks A and B, in such a way that if subnetwork A attempts to assign a class A_i to a binary input pattern I_A the result is an inference that subnetwork B will assign its simultaneously-occurring input I_B to a class B_j. The inference is the result of a strong connection between the F2 nodes for classes A_i and B_j in the two subnetworks, and this is denoted $A_i \rightarrow B_j$. In LAPART, each inference is tested in subnetwork B through its own vigilance pattern matching operation; if the subnetwork B vigilance system is not aroused (hence, the match of I_B to the B_j template pattern is accepted), the inference and, therefore, the subnetwork A classification decision were valid. Otherwise, the subnetwork A decision is assumed invalid. This is where the LAPART logic must adapt: The subnetwork A class templates must be modified appropriately if a mistaken inference is to result in a lasting correction. On the other hand, the inferencing connections between A and B classes, once formed, are assumed always correct. Further, they are assumed exclusive: Each subnetwork A class can infer only a single subnetwork B class. The phenomena that characterize the complexity of learning with LAPART-1 stem from these assumptions.

In Section 2, we briefly review the architecture and operation of ART-1, Stacknet, which converts real-valued input patterns to a binary structure, and LAPART-1. In Section 3 we present the LAPART-2 [11] algorithm as a means of addressing a phenomenon that can impede learning significantly in LAPART-1. Section 4 presents the learning theorems stating that a LAPART-2 network converges in two passes through a fixed training set. Along with this, we state a theorem showing that the architecture does not generate more templates than there are input examples. Section 5 describes the generalization study and its numerical results. Section 6 is the Discussion.

2   ART-1, Stacknet, and LAPART-1

In this section we briefly review the architecture and operation of ART-1, Stacknet, which converts real-valued input patterns to a binary structure, and LAPART-1.

2.1   Binary Patterns

First, we briefly review some notation and terminology. We shall regard a binary pattern X as a string of numerical 1s and 0s. Certain operations are defined upon binary patterns. First, if n is the number of 0-1 components, each denoted X_k, we write $length(X) = n$. For any two binary patterns X and Y having the same length, we refer to their component-wise minimum $X \wedge Y$, where the minimum operation on components has the properties, $0 \wedge 0 = 0, 1 \wedge 1 = 1, 0 \wedge 1 = 0, 1 \wedge 0 = 0$. For a set, S, of binary patterns all having the same length, with $S = \{X1, X2,...,XN\}$, let $\wedge S$ denote the pattern minimum over the set, $\wedge S = X1 \wedge X2 \wedge ... \wedge XN$. We define the size, $|X|$, of a binary pattern to be the number of 1s it contains. Finally we denote a "subset" relationship as $X \subseteq Y$, indicating that for every component in binary pattern X that has a 1 value, the same component in Y also has a 1 value.

2.2   ART-1 Architecture

To support our discussion of the LAPART architecture, we briefly summarize the function of an ART-1 network [8]. ART-1 is called an unsupervised learning architecture because it autonomously classifies its input patterns and "remembers" the classes in the form of binary connection-weight template patterns. An ART-1 network has three main layers of nodes. These layers consist of m_1 input (I) nodes, m_1 matching ($F1$) nodes, and m_2 classification (F2) nodes. The I layer serves as the network input interface, with each input node, I_k, supplying excitatory input to its corresponding $F1$ node, $F1_k$. Each binary input pattern I, where $length(I) = m_1$, specifies the activation values of the input nodes for the duration of the presentation of I as the current input. Thus, if input pattern component I_k has the binary value 1, then input node I_k has an activation value of 1 for that pattern, and 0 otherwise. Since the activation value I_k of each input node is directly transmitted to the corresponding node $F1_k$ through the $I_k \rightarrow F1_k$ connection, which has a fixed weight of unity, the initial pattern of activation values over $F1$ is identical with the input pattern. The $F1$ and $F2$ layers interact through adaptive connections, under the control of the gain control (GC) and vigilance (VIG) nodes. The template pattern for each class comprises the connection weights in the unique set of adaptive connections associated with an $F2$ node. At any time, each template, T_i for class i, corresponding to node $F2_i$ for $(1 \leq i \leq m_2)$, has the form

$$T_i = \wedge S, \qquad (1)$$

where S is the set of binary input patterns that has previously been assigned to the class corresponding to T_i and, consequently, may have contributed to the adaptive recoding of the template pattern. An input pattern may contribute to a template at one time and yet may become associated with a different template at a later time, as templates continue to undergo recoding. This effect will occur until the ART-1 network has *perfectly learned* its input space. The authors of the ART-1 architecture characterize the behavior of its unsupervised classification algorithm through stability results in [8]. Further results in [10] include a key learning theorem that states that if a fixed set of patterns is repeatedly presented to an ART-1 network, the algorithm will converge (i.e., perfect learning of the input set will occur) in a finite number of epochs, with the input patterns arbitrarily re-ordered on each epoch. Perfect learning means that each training pattern I in the set will have a maximal subset template T_i with $T_i \subseteq I$, where T_i is the largest such template $|T_i| \geq |T_{i'}|$ for all $T_i \subseteq I$. As a consequence, I will resonate directly with T_i; that is, I will be classified as a member of class i (this is called the *direct access property* [8]). Finally, no recoding of T_i will occur, since $T_i \subseteq I$.

Since the vigilance nodes of its ART-1 subnetworks play a fundamental role in the operation of a LAPART network, we review the role of a vigilance node. During the $F2$ competition following input of a binary pattern I, some $F2$ node, $F2_j$ say, wins the competition and tentatively becomes the exclusive class representative for I. However, if its associated template pattern T_j is such that

$$|I \wedge T_j| / |I| < \rho, \qquad (2)$$

where ρ is the ART-1 vigilance parameter, then the vigilance node, *VIG*, becomes activated. When this happens, a reset occurs over the $F2$ layer, and $F2_j$ becomes suppressed for the duration of the presentation of I. This eliminates $F2_j$ from the competition for representing I during the current input presentation. When no more resets occur, *resonance* is said to have occurred, and the input has finally been assigned a class. The ART-1 classification algorithm can be summarized as one that solves the combinatorial optimization problem stated as follows:

$$maximize \; |I \wedge T_\mu| / (\beta + |T_\mu|)$$
$$w.r.t. \; \mu$$

$$subject \; to \; |I \wedge T_\mu| / |I| \geq \rho \tag{3}$$

A solution value i for μ is the index of the F2 node $F2_i$ that represents the class assigned to I with associated template T_i.

For each ART-1 input pattern, unsupervised learning occurs in two phases: (1) recognition of the input pattern as a member of some class, and (2) updating of the class template through synaptic learning. During a resonance, the commonality of the input pattern and template is synaptically learned by the network by adapting the template weight values. This is expressed in the following binary pattern equation:

$$T_{i\text{-}new} = I \wedge T_{i\text{-}old} \tag{4}$$

which leads to the template property expressed in equation (1). When a class is first established, all connection-weight values in its template are *1s*. Many of these are changed to *0s* via the learning process as the network assigns input patterns to the class.

The next subsection presents a preprocessing network that converts a single real-valued input into a multicomponent pattern containing binary-valued components. The resulting coded pattern is well suited for the processing of an ART-1 network.

2.3   Stacknet

The neural network described in this sub-section, called Stacknet [6], transforms (codes) real-valued components into binary patterns that possess an important property vis-à-vis the processing that occurs within ART-1 networks: binary patterns that are "similar" in an ART-1 sense correspond to real values that are similar in magnitude. This is not true of the usual binary-coded-decimal format used in digital computers, in which 0 and 1 are coefficients of powers of 2. The codings used here are referred to as *stack numerals* and are similar to "thermometer codes" where a real number is mapped into an interval defined by real-valued *minimum* and *maximum* values. This interval is quantized into m subintervals, one of which contains the real input value. Associated with each subinterval is a logical variable. The stack numeral is constructed by setting all of the logical variables for subintervals less than or equal to the one containing the real-valued input to TRUE (or 1) and those above to FALSE or UNCERTAIN (0). If the interval is thought of as being a vertical structure, the set of logical variables forms a stack of 1s topped by 0s, totaling m components high. The precision of representation is set by the choice of the *max*, *min*, and m stack parameters and can be easily matched to the accuracy of a measured input value.

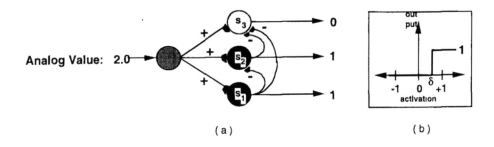

(a)   (b)

Figure 1. (a) A simple example of a neural implementation of a stack numeral. The complement stack has a different connectivity. (b) The activation function for the stack units with threshold δ.

A simple Stacknet is depicted in Figure 1. The connection strengths of stack inputs are all unity. Each stack node s_i $(1 \leq i \leq m)$ has an activation threshold of magnitude $\delta > 0$. Thus, stack node s_1 can be activated by a signal from the analog source node of magnitude δ. When activated, it emits a signal of strength unity through the connection to its corresponding ART-1 $F1$ node. Simultaneously, it emits a signal of strength unity through a system of $(m-1)$ inhibitory connections to higher stack nodes $s_2, s_3,\ldots s_m$. These connections each have weight δ, so that the connection-weighted inhibitory signal arriving at each target node above s_1 has strength $1 \cdot \delta = \delta$. The consequence of this is that an input signal of strength $x \geq \delta$ to the stack network from the analog source node is required to activate s_1. However, if $\delta \leq x < 2 \cdot \delta$, only s_1 will be activated, for the inhibitory weighted signal δ from s_1 arriving at each higher of the $m-1$ target nodes $s_2., s_3.,... s_m$ causes the total connection-weighted input t_t into target node s_i ($i > 1$) to be below threshold, that is

$$t_i = x - (1 \bullet \delta) < 2\delta - \delta = \delta$$

$$or \; t_i < \delta$$

which is below threshold, implying that all higher nodes will remain in their off state. Similarly, stack node s_2 sends out a set of $m-2$ inhibitory connections of strength δ to stack nodes $s_3., s_4.,... s_m..$ In general, stack node s_i inhibits the $m - i$ higher nodes. As a consequence, stack node s_i will be activated *if and only if* the input analog signal has strength $x \geq i \cdot \delta$. Thus, an analog number $\delta \leq x \leq (m+1) \cdot \delta$ in magnitude can be represented to within an absolute precision of magnitude δ by the Stacknet network. Thus, if $n \cdot \delta \leq x < (n + 1) \cdot \delta$ stack nodes $1, 2,...,n$ will be activated, producing the binary pattern

$$I = [\; 1111...1000...0]$$

where there are n $1s$ and $m-n$ $0s$. Stacknet takes a single real-valued input and produces a binary-valued output pattern of fixed length m.

Two stack numeral binary patterns are similar in the ART1 sense if and only if they fall into a class which is represented by the same template pattern. If the real-valued inputs were coded as powers of two, the usual representation on digital computers, this equivalence would not hold. For example, the numbers 127 and 128 represented in powers of two require 8 bits, with low-order binary digits to the right, yielding the patterns 01111111 and 10000000, respectively, with a difference in bits equaling 8 out of 8 total, or 100%. By contrast, if $\delta = 1$ a stack representation requires a minimum of 128 stack nodes (bits) to exactly code the numbers $1, 2, ..., 127, 128,$ yielding a difference in bits equaling 1 out of 128 total, or less than 1%. Stack numerals require more binary components but are more appropriate for coding numbers for ART-1 networks.

Now suppose that several Stacknets, each reading from a different analog source node, are arranged in an input array for an ART-1 system. Here, the total number of binary-valued components that will be generated will be $m = m_3 \cdot m_4$, where m_3 is the number of real-valued inputs to be represented and m_4 is the length of each output stack (assuming uniform stack size). Let $X = (x_1, x_2, ..., x_{m1})$ be an array of real-valued variables that are input to the array of Stacknets, and then $I = (I_1, I_2,...I_{m2})$ denotes the *concatenation* of binary stack outputs that represent the components of X to a precision δ, each of length m_4, so $m_1 = m_3 \cdot m_4$. The ART-1 network receives this composite pattern at its F1 layer.

Finally, as pointed out in the Introduction, ART-1 converges on a fixed training set of binary-valued input patterns in a number of presentation epochs equal to the number N of *different sizes* of input patterns, where the size of a binary-valued pattern is the number of *1*-valued components it possesses. If all input patterns have the same size, then only a single epoch is required. Stacknet has a variant that accomplishes this through the use of complement coding. If I is the normal "positive" binary-valued ART-1 input pattern of length m, then we can define the complement "negative" of this pattern to be a pattern I^c of length m, as $I^c = 1 - I$, where 1 is a pattern of all 1 components. By concatenating the positive and negative patterns together, we form a pattern $C = (I^c, I)$ of length $2m$. If $| I | = n$, then $| I^c | = m - n$. Therefore, $| C | = | I | + | I^c | = m$, independently of n. Using complement coding of stacks representing real-valued inputs will allow ART-1 learning to converge in a *single epoch* for any set of input data.

2.4   LAPART-1

The basic LAPART-1 network architecture [1] is based upon the lateral coupling of two ART-1 subnetworks, referred to as A and B. The interconnects between these two subnetworks force an interaction of the respective classifications performed by the ART-1 subnetworks on their inputs. This modifies their unsupervised learning properties to allow the learning of inferencing relationships between their respective input domains. This can be thought of as supervised learning, or supervised classification. In actuality, however, it is much more general. The usual sense of classification is that of creating a partition of the inputs, that is, separating them into disjoint sets, with a label (the desired output specified by the "teacher") attached to each element of the partition. With the LAPART architecture, we may actually label sets with *sets* – in other words, the network extracts rules with antecedent and consequent predicates. In this discussion, the sets in question will be referred to as classes, because they are sets labeled by ART-1 F2 nodes and coded in templates. Also, the inputs, ART layers, and templates will be labeled with an *A* or *B* referring to the *A* and *B* ART-1 subnetworks.

In a typical LAPART application, two ART-1 subnetworks are presented with a sequence of pairs of simultaneously-occurring input patterns IA_k and IB_k for subnetworks *A* and *B*, respectively. As *A* and *B* form class templates for their inputs, the LAPART-1 network learns inference relations between their classes by forming strong $F2A \rightarrow F2B$ interconnections between pairs of simultaneously-activated *F2A* and *F2B* nodes. Convergence of a LAPART network in a finite number of passes through a training set requires that it reach the following operational state: Presentation of any input pair *(IA, IB)* from the set shall result in pattern *IA* being immediately assigned a class in ART-1 subnetwork *A* through direct access to the class template. Through a strong, learned inferencing connection, the class *F2A* node shall signal a unique *F2B* node to which it is connected, forcing it to become activated. This results in the inferred *B* class template being read out over the *F1B* layer just as pattern *IB* reaches the *F1B* layer. The ensuing vigilance test in subnetwork *B* shall then confirm that the inferred class is an acceptable match for *IB*. That is, the network *B* vigilance node shall remain inactive. Further, the *B* class template shall be a subset template for *IB*. In summary, a final pass through the data shall result in no resets and no synaptic strength changes (i.e., no learning).

To show how a LAPART-1 network learns class-to-class inferences, or rules, from example input pairs, we give a brief summary of its algorithm. Initially, subnetworks *A* and *B* are untrained ART-1 networks. Their *F2* nodes are fully interconnected by $F2A \rightarrow F2B$ connections which are too weak to carry a signal of significant strength; that is, there are no learned inferences. As it processes each input pattern pair *(IA, IB)*, the LAPART network does one of two things: It

either forms a new rule or tests a previously learned one. It forms a new
rule exactly when subnetwork *A* forms a new class for its input *IA*. That
is, if *A* has no acceptable template pattern for *IA* it selects a previously
uncommitted *F2A* node, $F2A_i$ according to the ART-1 algorithm. Then,
it modifies the adaptive $FIA \rightarrow F2A_i$ and $F2A_i \rightarrow FIA$ connections so
that the newly committed template pattern TA_i equals the input *IA*. We
denote the newly initialized class by A_i. Following the selection of
$F2A_i$, meanwhile, subnetwork *B* has been allowed to read its input *IB*. It
selects a node $F2B_j$ and either initializes a new template TB_j or recodes
(modifies) a previously committed one. A subnetwork *B* class, B_j, has
now been selected simultaneously with the newly initialized subnet-
work *A* class A_i. Finally, the $F2A_i \rightarrow F2B_j$ connection strength increases
to a maximum, implementing an *inference relationship*, or rule, $A_i \rightarrow B_j$.
We write the rule in the form of an implication formula because the
future presentation of an input pair for which A_i is the resonating class
for the *A* input will result in the inference through the strong $F2A_i \rightarrow$
$F2B_j$ connection that class B_j is appropriate for the *B* input. This strong
connection will remain the sole one from A_i to a subnetwork *B* class
node.

If subnetwork *A* already contains a class template TA_i that resonates
with *IA* on the other hand, then it also has a previously learned class-to-
class inference relationship $A_i \rightarrow B_j$. Thus, $F2A_i$ primes $F2B_j$ through
the strong $F2A_i \rightarrow F2B_j$ connection, forcing $F2B_j$ to become active and
read out the class B_j template over the *FIB* layer. Thus, when it is
allowed to read its input, *IB*, subnetwork *B* performs its vigilance
pattern-matching test using the template pattern TB_j instead of one that
would have been selected through the ART-1 winner-take-all
competition in layer *F2B*. This is where the LAPART network tests an
existing rule: If the pattern match between the inferred class template
TB_j and the input pattern *IB* is not acceptable, that is if

$$| IB \wedge TB_j | / | IB | < \rho B,$$

where ρB is the vigilance threshold for subnetwork B, then a reset
occurs in subnetwork *B* – the inference has been disconfirmed. Through
the fixed, strong connection $VIGB \rightarrow VIGA$ between the two vigilance
nodes, subnetwork *A* is subsequently forced to also undergo a reset,

which we call a *lateral reset*. A lateral reset overrides subnetwork A's autonomous, or unsupervised, classification decision and forces it to find an alternative class for its input. The entire process must then be repeated using the reduced set of nodes obtained by inactivating $F2A_i$ and therefore $F2B_j$. Finally, the network either forms a new rule or modifies the templates that are linked through a pre-existing one.

It is interesting to ask whether the LAPART-1 algorithm always converges to the state in which no more resets or template modifications occur – all inferences are correct and learning has ceased. If it does not, are there conditions that can be specified under which it can be guaranteed to converge? Is there, at least, a set of well-defined necessary conditions for convergence? Unfortunately, it cannot be guaranteed that a LAPART-1 network will reach an operational state of convergence on the training set. Our attempts at addressing this issue resulted in the design of the LAPART-2 network and proofs of theorems stating that a LAPART-2 network converges in two passes through a training set.

3 The LAPART-2 Algorithm

In this section, we describe the LAPART-2 architecture [11], which implements neural network design constraints that we derived in order to resolve issues with LAPART-1. The LAPART-2 architecture is identical with the LAPART-1 architecture except in the procedure for a lateral reset. The modified lateral reset procedure results in a rule extraction neural network that converges in two passes through a set of training data, given that certain sufficient conditions hold for the data. Two-pass *supervised* learning is a special case of this, since, as mentioned before, rule consequents in supervised learning are simply class labels assigned by the teacher.

3.1 Forcing Learning to Occur

In LAPART-1, a lateral reset merely disqualifies the active $F2A$ node, forcing ART-1 subnetwork A to select an alternative resonant node from the set of all $F2A$ nodes that have not yet undergone a reset for the current input pair. As in the example let this pair be *(IA, IB)*. It can happen that *IA* has a direct access template TA_i whose choice results in a lateral reset, while a subsequently chosen template $TA_{i'}$ results instead in a valid inference; yet the latter template is also a subset template for *IA* (necessarily, it is smaller, having fewer binary *1* components). That no learning can occur in subnetwork *A* as a consequence of this (because only subset templates were chosen) means that the originally chosen template can remain the direct access template for *IA* afterwards. This allows the same sequence of events in subsequent passes to be repeated for the pair *(IA, IB)*, ensuring that the lateral reset (signaling an incorrect inference) will be repeated.

The LAPART-2 learning algorithm overcomes this learning deficit by allowing the choice of only an *uncommitted F2A* node to represent *IA* following a lateral reset. As a consequence, learning will occur, and in two forms. First, the uncommitted template will be re-coded as *IA*. This recoding represents the network's current state of knowledge about the new class, which consists of a single example. Since there is no $A \rightarrow B$ inference generated, the procedure for adding a new *A* class, $A_{i'}$, comes into play; subnetwork *B* produces a class template $TB_{j'}$ that resonates with *IB* in the usual ART-1 fashion. Unless this is a subset template for *IB* or else corresponds to an uncommitted *F2B* node, an existing *B*-class template will be modified. The second form of learning that occurs is the learning of a strong connection $F2A_{i'} \rightarrow F2B_{j'}$, which implements a newly learned inference relationship, $A_{i'} \rightarrow B_{j'}$.

3.2 Constraints on the Input Data

We shall state two learning theorems in Section 4, the most important result being the convergence of LAPART-2 in two passes through a fixed set of input pattern pairs. See [11] for an additional theorem. Unfortunately, the algorithmic modifications leading to the LAPART-2 architecture are insufficient, by themselves, for a proof of convergence. For this reason, the hypotheses of the learning theorems state assumptions that apply to the input data pattern pairs. Let $mA1$ and $mB1$ be the number of input pattern components IA_k and IB_l in the inputs to subnetworks *A* and *B*, respectively, and let *KA* and *KB* be integers such that $0 < KA < mA1$ and $0 < KB < mB1$. Hypothesis (i) is the statement that the input patterns for each ART-1 subnetwork have a fixed size, *KA* for subnetwork *A* and *KB* for subnetwork *B*. This may appear to be a strong constraint. However, it is less strong an assumption than is routinely applied with ARTMAP and Fuzzy ARTMAP [4], [5]: In applications of these architectures, it is normally assumed that each input pattern is complement coded, with the effect of making all input patterns the same size.

Hypothesis (ii) in the Two-pass Learning theorem is more complex: It is meant to ensure that LAPART-2 is a *consistent learner* (see [13]). A consistent learner is a machine which, given consistent training data, can successfully learn some specified target concept from that data. In the learning of class-to-class inferences (rule extraction), we apply the assumption that the input pattern pairs are consistent and, as a result, are able to prove that LAPART-2 converges. In the context of LAPART, consistency means that the pattern minimum ($\wedge$) of the subnetwork B input patterns with which each subnetwork A input pattern is paired can form a template with which each one of them would pass the vigilance test. Without this hypothesis, there could be a subnetwork A input pattern IA for which the LAPART network was incapable of learning correct B inferences; there would always be some B input pattern IB associated with IA that would cause a lateral reset.

4   The Learning Theorems

We can now state the learning theorems for the LAPART-2 neural network architecture. See [11] for the details of the proofs. Only the hypotheses pertaining to the data are stated explicitly. The neural network behavioral hypotheses are implicit in the statement in the theorems. In the following, let L be a LAPART-2 network with mA and mB input nodes for subnetworks A and B and with vigilance values ρA and ρB. Let MA and MB be sets of input patterns for subnetworks A and B, respectively, and let M be a set of input pattern pairs $(IA_k, IB_{k,h})$, with $IA_k \in MA$, $IB_{k,h} \in MB$ $(k = 1,...,N; h= 1,..., n_k)$. Finally, let $MB_k = \{ IB_{k,h} \mid h = 1, ..., n_k \}$. The first theorem follows:

<u>Theorem (Two-pass Learning)</u> Let L be a LAPART-2 network whose inputs have the following two properties:

(i) $\mid IA_k \mid = KA$, $\mid IB_{k,h} \mid = KB$ $(0 < KA < mA; 0 < KB < mB)$.

(ii) For an arbitrary subset $S \subseteq MB$ and for any $IA_k \in MA$, if an associated pattern $IB_{k,h}$ is in S (i.e., if $IB_{k,h} \in S \cap MB_k$)

and $| \wedge S | \geq \rho B \cdot KB$

then $| (\wedge S) \wedge (\wedge MB_k) | \geq \rho B \cdot KB$.

Then, if each of the elements of M is input to L in each of several passes, with the elements arbitrarily ordered in each pass, there will be no resets and no new templates in subnetworks A and B, and no changes in class assignments in subnetwork A, following the second pass. Recoding can occur only in subnetwork B templates following the second pass. Any such recoding will occur only on the third pass and will have no effect upon the class assignments and inference relationships that L has learned in the first two passes.

Although hypothesis (i) is essential, it is also one that is commonly applied in studies of ARTMAP and LAPART type architectures, and is even considered essential for the correct performance of ARTMAP [4], [5]. It is hypothesis (ii), together with the LAPART-2 modification itself, that is uniquely responsible for the two-step convergence result. This hypothesis, however, is difficult to verify for a given application. It specifies that *any* template that could conceivably be associated with the subnetwork B input patterns that are paired with a single subnetwork A input pattern must admit all of them. For the intended rule extraction applications of LAPART, in which sets of A inputs (rule antecedents) are to be associated with sets of B inputs (rule consequents), it would be impractical to check this. Also, the condition is rather strong – probably stronger than necessary – and is not likely to hold in some cases. See [11] for a further modification of the LAPART architecture that addresses this issue. In the csae of pure clasification problems, like those presented in Section 5, hypothesis (ii) may easily be shown to hold true.

Theorem (LAPART Data Compression) Let L be a LAPART network which is processing input pattern pairs $(IA_k, IB_{k,h})$ from a set M. Suppose that the B inputs all have the same size $| IB_{k,h} | = KB$ for all applicable values k, h. Then, the number of laterally connected template pairs

(TA$_i$, TB$_j$) generated by *L* does not exceed the number of input pairs (*IA$_k$, IB$_{k,h}$*) in *M*.

Notice that the LAPART Data Compression theorem requires no constraint on the architecture – it can be any of LAPART-1 or 2 variants. The only restriction on the input data is that the *B* component of all input pattern pairs be the same size. This is much weaker than the hypotheses in the Two-pass Learning theorem. A consequence of the LAPART Data Compression theorem is that the number of $A_i \rightarrow B_j$ rules extracted can be no greater than the number of input data pairs. Neither template proliferation nor rule proliferation is a problem with a LAPART network. The following section further explores the properties of LAPART-2 architectures through numerical simulations.

5   Numerical Experiments

With most learning systems, it is frequently possible to achieve near perfect learning on a fixed training set of data at the expense of either using a large enough set of synaptic weights in the network or reduced performance on an independent testing set of data. The former effect is addressed by the template proliferation result mentioned above. The latter effect is referred to as poor generalization or over-training. Since a theoretical understanding of generalization in LAPART class architectures is still under development, this section addresses the topic through a series of numerical experiments on challenging problems in classification. Note that this class of problem has been used in these studies because of the simplicity of their correctness analysis and the availability of independent theoretical bounds on performance. Note also that issues in generalization exist equally in non-classification type problems, such as inference and rule learning [2], [3].

5.1   Method

Three classification problems were selected to study generalization in LAPART-2 learning [12]. Each problem has the properties of being a two-class problem, with two real-valued feature-space dimensions (x0, x1) for input into the A subnetwork, with statistical overlap between the two class boundaries, and the ability to generate the data ordered pairs algorithmically. The input variables are confined to the [0,1] interval. The three study problems are:

1) two equal sized rectangular uniformly distributed classes with 50% overlap in the x0 dimension,

2) two overlapping normally distributed classes each with different means of (0.333, 0.5) and (0.666, 0.5) respectively, and the same sigmas (0.166, 0.166),

3) two overlapping normally distributed classes each with the same means of (0.5, 0.5) and differing sigmas (0.166, 0.166) and (0.333, 0.333) respectively.

A computer simulation of LAPART-2 was used to experiment with the three study problems. For each problem, training and testing data sets were independently created using a numerical random number generator that modeled the appropriate statistical distribution. A total of 1000 ordered pairs were produced for a data set (training and testing), 500 for Class 1 and 500 for Class2.

LAPART-2 was configured using complement coded stack (CCS) representations for inputs to both the A and B subnetworks [6]. The input to subnetwork A consisted of two concatenated CCS representations, one for each input dimension, using 1024 bits in the positive stack. The input to unit B was a single CCS representation using 2 bits in the positive stack. The two classes were labeled 10 and 01 respectively.

An experiment consisted of training a LAPART-2 network on the training set until convergence, then computing a performance measure using the testing data set with learning disabled. Since the details of learning in this class of network depends on presentation order of the training data, the performance measures from training with twenty different random orderings were averaged and standard deviations were computed. In addition, statistics for the number of learned templates in the A subnetwork was collected. This gives an indication of the degree of data compression realized by the network. Convergence was declared for a training session when at the end of a presentation epoch, each training pattern had a direct access template [8] in both the A and

B subnetworks. Notice that this requirement is more demanding than is required for the conclusion of the Two-pass Learning theorem.

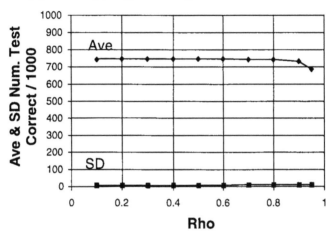

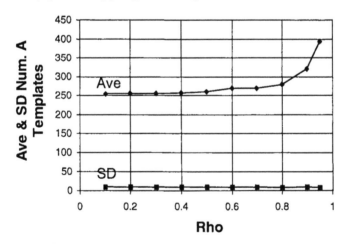

Figure 2. Overlapping Rectangular Distributions: (a) the average and standard deviation for the number of correctly classified testing samples out of 1000 as a function of rho for the A subnetwork; (b) the average and standard deviation for the number of A unit templates as a function of rho.

Since learning in ART-class architectures is also dependent upon the vigilance parameter [8], ρ, the average performance was computed on a grid of ten vigilance settings (0.1, 0.2,...,0.9, 0.95) for the A subnetwork. The vigilance setting for the B subnetwork was fixed at 1.0. This is standard for classification problems, since binary coded class labels are used as inputs to the B subnetwork. Finally, a Bayesian classifier was applied to the testing data and performance was calculated, giving a basis for comparison.

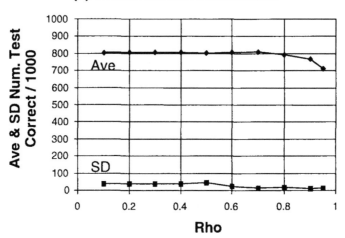

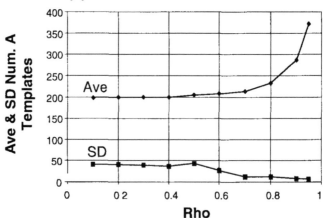

Figure 3. Offset Normal Distributions: (a) the average and standard deviation for the number of correctly classified testing samples out of 1000 as a function of rho for the A subnetwork; (b) the average and standard deviation for the number of A unit templates as a function of rho.

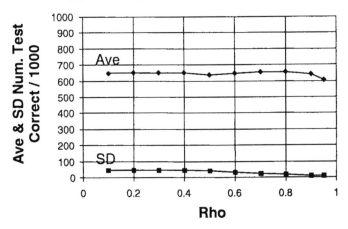

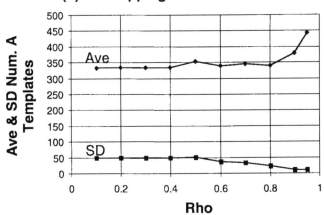

Figure 4.   Aligned Normal Distributions: (a) the average and standard deviation for the number of correctly classified testing samples out of 1000 as a function of rho for the A subnetwork; (b) the average and standard deviation for the number of A unit templates as a function of rho.

5.2   Results

The averages and standard deviations for the number of correctly classified testing data set members are given for the three problems as a function of A subnetwork ρ value in Figures 2a, 3a, and 4a. The averages and standard deviations for the number of A subnetwork templates are give in Figures 2b, 3b, and 4b. Table 1 gives a summary of the LAPART-2 performance results, including Bayesian performance for comparison.

Table 1. Summary of Performance Results for A subnetwork Rho=0.1. The numbers in parentheses are the standard deviations resulting from the averaging of 20 different orderings of the training data set. "Performance" measures the average percentage correct classification on the independent testing data sets.

A Rho = 0.1	Problem 1 (Rect)	Problem 2 (Norm)	Problem 3 (Norm)
Training Epochs	1.4 (0.5)	1.3 (0.45)	1.8 (0.40)
# A Templates	260 (10)	200 (40)	335 (50)
Bayesian Perf	~75%	~84%	~73%
LAPART Perf	75% (1%)	81% (4%)	65% (5%)

6   Discussion

The testing data set performance of LAPART-2 closely matches that of a Bayesian classifier for each of the three problems. A lower average accuracy is to be expected, given that we are applying a nonstatistical method to a problem defined in terms of statistical information. Note that the performance varies very little with respect to the A subnetwork vigilance (ρ) over wide ranges of the parameter, and that performance is generally better at lower values. This is partially due to the larger maximum hyperbox size allowed by smaller ρ values, resulting in greater loss of binary 1s in the template patterns formed using complement-coded stack input patterns [2].

Note also that convergence occurs on the average in less than two epochs, as predicted by the Two-Pass Learning theorem stated in a previous section. In many cases, only a single epoch was required to perfectly learn the training data.

One important question deals with the ratio of the number of learned A subnetwork templates to the total number of training samples. If this ratio is near 1, it would indicate a high degree of pattern memorization. This is usually a predictor of poor generalization performance. However, LAPART-2 demonstrated a ratio of around 0.25. This indicates that very little memorization is occurring, consistent with the good testing performance data. Some memorization is to be expected given the propensity of LAPART-2 to create templates accessed by only one training pattern [11]. Note that because of the use of stack input representations, a "point hyperbox" is not really a point – it codes a small region of feature space within a stack interval.

7 Conclusion

LAPART-2 has a distinct advantage over LAPART-1 that stems from the modification that forces learning to occur in response to each lateral reset. We have stated that a LAPART-2 network, given the assumptions upon the input data that we described, converges in two passes through a set of training data, with the pattern pairs arbitrarily ordered on each pass. The convergence bound for ARTMAP is greater, varying with the size $mA1$ of the binary input space for subnetwork A and with its vigilance value ρB [4]. Finally, in [11], we proved that template proliferation does not occur despite the requirement that a new subnetwork A class be initialized with each lateral reset.

Our results are especially significant in that they apply to rule extraction with a network that partitions its input and output spaces (A and B) into classes, as opposed to simple class labeling. Thus, each subnetwork A input can be associated with many subnetwork B inputs. This allows for the learning of rules as class-to-class inference relationships as well as inferencing under uncertainty.

The numerical studies presented in this chapter demonstrate that LAPART-2 has one of the tightest bounds known on learning convergence. Additionally, they provide empirical evidence that this need not compromise generalization performance. These results have many implications for the utility of this architecture in future application domains.

Acknowledgements

We would like to acknowledge the support of the University of New Mexico Albuquerque High Performance Computing Center, the University of New Mexico Department of Electrical and Computer Engineering, and The Boeing Company.

References

[1] Healy, M.J., Caudell, T.P., and Smith, S.D.G. (1993), "A neural architecture for pattern sequence verification through inferencing," *IEEE Transactions on Neural Networks*, Vol. 4, No. 1, pp. 9-20, January.

[2] Healy, M.J. and Caudell, T.P. (1997), "Acquiring rule sets as a product of learning in a logical neural architecture," *IEEE Trans. on Neural Networks*, Vol. 8, pp. 461-475.

[3] Caudell, T.P. and Healy, M.J. (1996), "Studies of inference rule creation using LAPART," presented at the IEEE Conference on Neural Networks, Washington, D.C., (ICNN`96). Published in the *Proceedings of the Fifth IEEE International Conference on Fuzzy Systems, (FUZZ-IEEE)*, New Orleans, LA, pp. ICNN 1-6.

[4] Georgiopoulos, M., Huang, J., and Heileman, G.L. (1994), "Properties of learning in ARTMAP," *Neural Networks*, Vol. 7, No. 3, pp. 495-506.

[5] Carpenter, G.A., Grossberg, S., Markuzon, N., Reynolds, J.H., and Rosen, D.B. (1992), "Fuzzy ARTMAP: a neural network architecture for incremental supervised learning of analog multidimensional maps," *IEEE Transactions on Neural Networks*, Vol. 3, pp. 698-713.

[6] Healy, M.J. and Caudell, T.P. (1993), "Discrete stack internal representations and fuzzy ART1," *Proceedings of the INNS World Congress on Neural Networks*, Portland, Vol. II, pp. 82-91, July.

[7] Healy, M.J. (1993), "On the semantics of neural networks," in Caudell, T.P. (Ed.), *Adaptive Neural Systems: The 1992 IR\&D Technical Report*, Technical Report BCS-CS-ACS-93-008, available from the author c/o The Boeing Company, PO Box 3707, 7L-66, Seattle, WA, 98124-2207.

[8] Carpenter, G.A and Grossberg, S. (1987), "A massively parallel architecture for a self organizing neural pattern recognition machine," *Computer Vision, Graphics, and Image Processing*, 37, pp. 54-115.

[9] Healy, M.J. (1999), "A topological semantics for rule extraction with neural networks," *Connection Science*, vol. 11, no. 1, pp. 91-113.

[10] Georgiopoulos, M., Heileman, G.L., and Huang, J. (1991), "Properties of learning related to pattern diversity in ART1," *Neural Networks*, Vol. 4, pp. 751-757.

[11] Healy, M.J. and Caudell, T.P. (1998), "Guaranteed two-pass convergence for supervised and inferential learning," *IEEE Trans. of Neural Networks*, Vol. 9, pp. 195-204.

[12] Caudell, T.P. and Healy, M.J. (1999), "Studies of generalizations for the LAPART-2 architecture," *Proceedings of the IJCNN*.

[13] Heilman, G.L., Georgiopoulos, M., Healy, M.J., and Verzi, S.J. (1997), "The generalization capabilities of ARTMAP," *Proceedings of the IJCNN*, Houston, TX.

CHAPTER 7

NEURAL NETWORK LEARNING IN A TRAVEL RESERVATION DOMAIN

H.A. Aboulenien and **P. De Wilde**
Department of Electrical and Electronic Engineering
Imperial College of Science, Technology and Medicine
London, U.K.
{h.aboulenien,p.dewilde}@ic.ac.uk

This chapter presents an intelligent agent that employs a machine learning technique in order to provide assistance to users dealing with a particular computer application. Machine learning is a sub-field of artificial intelligence (AI) that includes the automated acquisition of knowledge. The aim is intelligent systems that learn, that is, improve their performance as a result of experience. The agent learns how to assist the user by being trained by the user using hypothetical examples and receiving user feedback when it makes a wrong decision. The main aim is to reduce work for the end-user by building a software agent that acts as a personal assistant. The proposed interface agent will potentially have the features of natural language interface and learning through interaction. The achievement of these innovations is mainly based on neural network learning techniques. The chapter presents preliminary results from a prototype agent built using this technique and applied on flight reservation domain.

1 Introduction

One of the obvious difficulties for building intelligent machines has been for many years the passive nature of computers. Computers do only what they were programmed to do and do not learn to adapt to changing circumstances. At the same time, it will become more and more difficult for untrained computer users to cope with the increasing complexity of computer applications and the growth of the computers' direct-manipulation interfaces. One of the aims for building intelligent machines, and also one of the biggest challenges, is to create a simple user interface so that the human-computer interaction will become as natural for end-users as picking up a phone or reading a newspaper.

Artificial intelligence researchers and software companies have set high hopes on so-called *Software Agents* which learn users' interests and can act autonomously on their behalf to contribute in solving the learning problem [1]. The learning approach has several advantages. First, it requires less work from the end-user and application developers. Second, the agent can easily adapt to the user over time and become customised to individual and organisational preferences and habits. Despite the huge diversity in this rapidly evolving area of agents' research, the most promising one in solving the human-computer interaction problems is called *Interface Agents*. Interface agents can be characterised as systems, which employ artificial intelligence techniques to provide assistance to users with a particular computer application [2].

An important property that any interface agent should have is the ability to communicate with the human user via some kind of natural language. This is based on the belief that computers will not be able to perform many of the tasks people do every day until they, too, share the ability to use their language. Despite more than twenty years of research into natural language understanding, the solutions for actual problems such as a natural language computer interface still suffer from inadequate performance. This problem is due to the fact that these systems depend on exact knowledge of how human language works. However, even now there is no complete and formal description of human language available. It has been argued that there is some evidence that the human brain is specially structured for language [3]. However, today's computer architecture is totally different from the human brain.

In an attempt to model the human mind/brain, it has been necessary to oversimplify the structure and the function. This has led to the development of an important area of research, namely neural computing. This area belongs to a larger research paradigm known as computational intelligence which aims to model functions associated with intelligence, at the signal level as a dynamical system. Neural computing is the study of Artificial Neural Networks (ANNs).

In this chapter, the main aim is to introduce the interface part of the proposed agent that communicates with the user and learns through this interaction to be able to assist him/her in the travel reservation domain. The learning machine of this agent is based on neural network techniques. In the next section, we introduce a brief definition of agents. Section 3 presents the main features of neural networks. The rest of the chapter explains the implementation of the proposed interface agent and some of the simulation results.

2 Agents

Software agents have evolved from multi-agent systems, which in turn form one of three broad areas, which fall under distributed artificial intelligence (the other two being distributed problem solving and parallel artificial intelligence). Although the term agent is a widely used term in computing, AI and other related areas, it is poorly defined. Perhaps the most general way in which the term agent is used is to denote a hardware or (more usually) software-based computer system that enjoys the following: [4]

- *Autonomy*: Agents operate without the direct intervention of humans or other agents and have some kind of control over their own actions and internal state.
- *Social Ability*: Agents interact with other agents (and possibly humans) via some kind of agent-communication language.
- *Reactivity*: Agents perceive their environment (which may be the physical world, a user via a graphical user interface, a collection of other agents, the INTERNET or perhaps all of these combined) and respond in a timely fashion to changes that occur in it.
- *Pro-activeness*: Agents do not simply act in response to their environment; they are able to exhibit goal-directed behaviour by taking the initiative.

There are sometimes other agent's attributes, which are considered secondary attributes to those mentioned above such as *mobility, continuity, robustness* and *rationality*. Our proposed interface agent aimed to enjoy most of the mentioned attributes except the mobility feature. Generally, the implementation of interface agents is focused on **autonomy and learning**. **The interactivity (social ability) attribute will** be gained through the architecture of integrating many agents in a collaborative scheme.

3   Neural Network Role

Our aim was to benefit from the features of artificial neural networks (ANNs), which mimic the biological nervous system to perform information processing and learning. On top of the superficial resemblance, ANNs exhibit a surprising number of human brain characteristics such as learning, generalisation and robustness. One of the most important features of ANNs is the ease with which they can learn (modify the behaviour in response to the environment). *Learning* in ANNs is the process of adjusting the connection weights between the nodes. Neural networks are often required to learn an input/output mapping from existing data or learn from input data only when the output is not known. In the last case, ANNs are capable of *abstracting* the essence of a set of input data, i.e., learning to produce something never seen before. ANNs perform the equivalent of inductive learning in the symbolic paradigm.

Once trained, a network's response can be, to a degree, insensitive to minor variations in its inputs. *Generalisation* in learning enables ANNs to learn from incomplete data. This ability to see through noise and distortion to the pattern that lies within is vital to pattern recognition in a real-world environment. Producing a system that can deal with the imperfect world in which we live overcomes the literal mindedness of the conventional computer. This attribute is mainly due to its structure, not by using human intelligence, embedded in the form of *ad hoc* computer programs. Computer programs play no role. *Parallelism* allows ANNs to model complex relations and to perform complex tasks at speeds in excess of other algorithms (this feature can only be fully exploited if their hardware implementation is used). The above features are the main contribution of ANNs to intelligent systems [5].

As a matter of fact, ANNs have proved to complement conventional symbolic artificial intelligent techniques in applications where the theory is poor and the data are rich, such as pattern recognition, pattern matching, and adaptive non-linear control. Some researchers claim that ANNs will replace current AI, but there are many indications that the two will co-exist and be combined into systems in which each technique performs the tasks for which it is suited [6].

4   Agent Architecture

In this section, we introduce a brief description of the proposed interface agent's building blocks. Figure 1 shows the interface agent architecture [7].

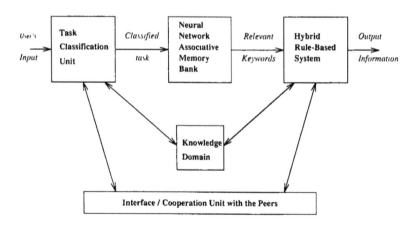

Figure 1. The interface agent architecture.

- **Task Classification Unit (TCU):** This is the interface part where the user-system interactions are performed. The TCU accepts the input request as natural language sentences from the user. The sentences represent the required task that is required to deal with. TCU applies approximate string matching and neural network learning techniques to analyse the input sentences, then classifies the task according to a search process in its heuristic database system (concept maps).
- **Neural Network Associative Memory (NNAM):** This part of the agent acts as a heteroassociative memory which is responsible for

generating keywords related to the classified task received from the TCU, though it works as a look-up table implemented using ANN technique.

- **Hybrid Rule-Based System (HRBS)**: HRBS applies simple rules on the agent's knowledge domain, using the keywords, which are generated from NNAM, to infer the actual information required in executing the user's task.
- **Peers Co-operation Unit (PCU)**: It is a channel to cooperate with other agents to ask and answer questions.

In the rest of this chapter, the discussion is mainly concentrated on the implementation issues of the task classification unit, in which the neural network learning techniques are applied.

4.1   Problem Domain

Travel reservation is considered as a problem domain to apply the ideas of agent-user interaction through natural language interface and learning over time in order to be able to assist the user. The aim is to build a task classification unit that is able to classify the user's input request to a specific task category.

By natural language interface is meant manipulating a short sentence or phrase and allowing misspelling and/or ungrammatical cases. We do not claim that the agent deals with this problem in a satisfactory way from the semantic or lexical analysis point of view. Since a deep text analysis cannot be undertaken in an unrestricted semantic environment, the approach must be to limit the task in order to analyse the user's input text as well as possible. It has been claimed that most of the successful natural language understanding systems share two properties: they are focused on a particular domain rather than allowing discussion of any arbitrary topic, and they are focused on a particular task rather than attempting to understand language completely [8].

In order to be able to assist users, an agent must be provided with knowledge of its domain. The agent is given a minimum of background knowledge and learns the appropriate behaviour either from the user or from other agents. By learning is meant learning by example and accumulating the experience from interacting with and observing its user.

To tackle the problems of learning and natural language interface, it is assumed that the agent's vocabulary is restricted to the task domain language, i.e., the whole language is divided into many sub-languages. Later, the results will show that the neural network approach is able to associate the user's input sentence (information as a stream of words) with the user's concept within this restricted domain. This assumption enables us to design a simple user-agent interface form to accept unrestricted simple natural sentences or phrases.

4.2   Data

Twenty-five e-mails written in the English language are collected from persons with different native languages. Each e-mail contains a few sentences or phrases representing the three specified categories in the airline travel reservation area. These three categories are:

1. Asking about travel reservation.
2. Asking about travel confirmation.
3. Asking about travel cancellation.

It has been asked that every respondent write a short sentence or a phrase representing as much as possible of the meaning without any concern about the correct grammar. The e-mails contain a mix of formal and informal English language. The only correction, which has been made before this set of data has been applied on the neural network for training, was a spelling check. In this approach, it is assumed that:

*   The user's sentences (most of the input stream of words) are within the vocabulary domains.
*   The sentences are not differentiated according to their grammar.
*   The user's input sentences are not compound (simple requests within the domain language).

The collected e-mails (dataset) consist of a combination of more than one hundred sentences and phrases. This dataset contains more than three hundred different words. A few no-meaning phrases are added to represent the neutral category in the classification process. The neutral category is supposed to include all the common words that have no effect on the task identification process (reservation, confirmation or cancellation) such as a country name. Part of the dataset is chosen as a training group (approximately 30% of the whole dataset). All the sentences and phrases are contained in the test group.

Two types of ANN architectures have been trained using this dataset: a Multi-Layer Feedforward Network (MLF) and a Single-Layer Perceptron Network (SLP). With certain adjustments to the initial weights, learning constant and steepness coefficient, both neural network architectures give the same result. Changing the input representation from unipolar (where active neurons are represented by +1 and 0 represents in-active neurons) to bipolar (where active neurons are represented by +1 and in-active neurons are represented by -1) has a slight effect on the network performance. The activation of a neuron depends on the presence of the word that the neuron represents in the input stream of words. The order of the training data has more influence on the training process. This is due to the finite size of the training set. Here is an example to explain what training order effect means. The word Egypt has the following weights according to the training set in which the word is encountered four times (one time per category).

	reservation	confirmation	cancellation	neutral
Egypt	-4.414566	-8.017722	-0.757465	6.551010

However, the ideal weights for such a word should be approximately the same for all categories but the neutral category. In the actual case, there are differences because of the finite size and the order of the training set (order means which category was trained first). If the words appeared many times in a random category order, this problem could be overcome. On the other hand, this will elongate the training time. It has been deduced that, in order to obtain a fair word-to-concept map distribution from any trained neural network, the training dataset must be carefully selected. In other words, the training sentences and phrases should be prepared such that the neutral (common) words must be represented equally for all the defined categories in the training dataset. This is the main reason for adding the no-meaning phrases: to teach the neural network how to neutralise the effect of the presence (absence) of these words in the user's input sentence/phrase. The neutral category contains all the words which should have equal weights towards the three categories mentioned above such as London, to, on, for, ... and/or from.

The other factor which affects the learning time is the training set size. The two architectures are trained with different size training sets. It is obvious that the larger the training set the longer the training time. In the next section, we will explain two learning processes which complement each other to compensate the above mentioned pitfalls in the training set.

4.3   Network Training

There are two types of training for either an MLF network or an SLP network: an off-line training and an on-line training. The off-line training aim is to construct concept maps from the training examples. These concept maps relate the each input sentence/phrase to a specific concept in the problem domain. The off-line training takes place before all the operations in order to assign connection weights to all the words in the training set patterns. A pattern consists of a unipolar (bipolar) representation of the training sentence or phrase. For example, the sentence could be:

Due to unforeseen circumstances I have to cancel my flight to London

Then the pattern would be:

-1 -1 -1 -1 -1 -1 -1 1 -1 ... -1 -1 1 -1 1 -1 -1 -1 1 -1 -1 -1 -1 -1 1 1 -1....

Each input neuron represents a word and each output neuron represents a category. The learning is according to the generalised Delta learning rule: [9]

$$c(d_i - o_i) f'(net_i) y_j \qquad\qquad j = 1, 2, \ldots, n$$

$$net_i = \sum_{j=1}^{J} w_{ij} y_j$$

$$o_i = f(net_i)$$

where

w_{ij} is the weight that connects the output of the jth neuron with the input to the ith neuron

o is the output vector

c is the scaling factor

f is the activation function (sigmoid function)

f' is the activation function first derivative

y is the input vector

d is the target vector

Figure 2 is the graphical representation of the classified words into the correct category after the off-line training is completed.

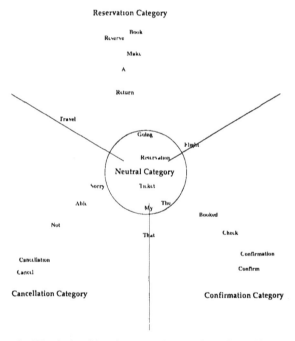

Figure 2. Word classification sample based on the collected data.

Table 1 shows a sample of the weights for some words after SLP[1] network training that takes place based on the collected dataset. Before the training all the words were assigned small random weights.

---

[1] SLP results show a clearer relation between words and categories than MLF.

Table 1. SLP training results (sample).

WORD	CATEGORIES			
	RESERVATION	*CONFIRMATION*	*CANCELLATION*	*NEUTRAL*
a*	8.160717	-6.392762	-9.916912	-11.262650
able	-3.469685	-1.873571	7.569377	-4.008598
airline	1.778800	-6.574209	-2.470539	-7.258359
belgium	-0.636226	-6.133667	-3.362907	3.883109
book	26.382210	-13.742558	-8.563011	-7.614223
booked	-2.502439	8.311028	-2.922199	-3.853302
cancel	-21.769215	-24.374889	53.473717	-20.940538
cancellation	-21.756435	-12.751144	31.506802	-2.847031
check	-14.144594	22.119167	-2.908264	-2.108434
confirm	-16.029274	34.444370	-20.502920	-8.255243
confirmation	-21.811161	13.502781	-17.121601	-5.858447
egypt	-4.414566	-8.017722	-0.757465	6.551010
flight	2.957987	3.815885	-7.900447	-8.968528
going	5.197784	-8.156089	-7.538003	5.770500
make	11.682374	-6.972619	-1.942448	-14.292900
my	-5.785409	2.654307	1.096750	3.675545
need	-3.990644	-0.253858	-8.594911	5.843002
not	-3.469685	-1.873571	7.569377	-4.008598
reservation	1.135525	2.715665	1.451366	0.494692
reserve	20.650588	-4.416985	-2.991275	-8.332147
return	6.652804	-1.805905	-2.937553	-0.380169
send*	-0.74305	3.753846	-3.041971	3.572052
sorry	-6.61513	-5.857497	2.198464	-0.346198
the*	-14.7968	2.616945	-0.955008	2.513896
ticket	1.48133	5.895713	2.516594	0.510546
travel	8.168785	-7.046576	1.490917	-8.466378
trip	-7.755359	-6.876982	-0.422680	5.903522
want	-4.800109	-5.958756	-3.849468	6.855282

Inspection of the above results indicates that there is a strong connection between some words and a specific category. For example: words like **book, reserve** and **make** are directly related to the *reservation* category, while words like **check** and **confirm** are connected to the *confirmation* category. On the other hand, words like **reservation** and **ticket** are common among more than two categories. We intentionally added words like **send** to the training set to illustrate the effect of the unequal representation of the neutral words. The word **send** has appeared only twice in the training dataset: one time in a confirmation training example and the second time within the no-meaning phrases. It can be seen from the table that this word is related to both *confirmation* and neutral categories; however, it is supposed to be in the *neutral* category only. Finally, it can be noticed that some words like **the** and **a** are assigned into a certain category and this is interpreted as language dependent. In other words, most of the users use **a** when they ask about a ticket reservation and they use **the** when they ask about the ticket confirmation.

The other learning process is called the on-line training. The on-line training takes place during the user-agent interaction cycle. The network adapts some of its weights according to the user's direct response in order to correct the misclassified sentence/phrase to a different category. Also, it assigns weights for new words. The on-line learning rule is:

$$\mu_{ij}\varepsilon(d_i - o_i)$$

where
μ_{ij} is a multiplicative factor
ε is an on-line learning coefficient
d, o as defined above in the off-line training.

Each word has been assigned a correctness level value which is dependent on how many times the word is used before. The multiplicative factor is inversely related to the word correctness level value. Any word in the user's input and not belonging to the domain dictionary is considered new. The newer the word the higher the multiplicative factor and vice versa. The correctness level value of a word is increased each time the word is encountered in the input stream of words and the user's request is classified correctly. Hence, the change in the weights of the more often used words is much slower than the change in the weights of the new words during the on-line learning. The on-line learning coefficient is a small number defined by the user to control the speed of the on-line training process.

The on-line training is a complementary process to the off-line training. Also, it introduces a solution to the lack of enough data in the training set which leads to some sort of incorrect bias in the network weights. In addition, the whole learning process time is divided between those two processes.

5   Operation

Figure 3 shows the block diagram of the task classification unit in the interface agent. Once the off-line training process has been completed, the system is ready to move to inference and on-line training through agent-user interaction. The interaction cycle works as follows:

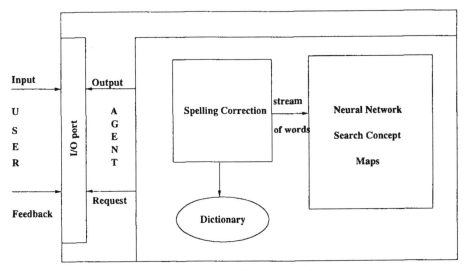

Figure 3. Task classification unit block diagram.

- **Step1**) *User-system interaction*: The user's information is provided as a simple sentence or phrase at the I/O port.
- **Step2**) *Approximate string matching*: Given the domain dictionary string set, the user's input stream of words is matched against the dictionary items to correct any spelling errors in the input words. A non-matching text item can be computed with a dictionary directly, using an iterative character-by-character matching scheme that determines the smallest number of single-character operations (insertions, deletions, substitutions, and transpositions) required to transform one of the words into the other. This is a well-known technique of approximate string matching [10]. It is assumed at this stage that the dictionary contains the travel domain vocabulary only. This means some correct English words might be considered misspelled and corrected according to this vocabulary domain. For example:

(Not in the dictionary)   **reverse → reserve**   (Nearest match)

The output of this step is the correct stream of words (if found) and the rest will be considered new words, even the wrong ones, until the system receives a feedback from the user in step5.

- **Step3**) *Pattern generation*: Some of the input neurons are activated according to the words presented in the user's input. The input neurons in the neural network are chosen large enough to accommodate the new words. The rest of the neurons that are not active indicate the absence of the corresponding words from the input stream of words. A new word can be assigned to any unused neuron or a neuron that represents the neutral word, which has the least correctness level value. This is the process of generating the patterns from the input stream of words. The generating patterns guide the neural network to an initial state (via a trajectory) in the concept maps that have been constructed during the off-line training phase.

- **Step4**) *Search process*: The neural network continues the search process to classify the input pattern to a certain category according to the off-line training. The output of this step is either directed to the I/O port to get a feedback from the user or is considered correct by the agent. This decision is mainly dependent on a factor defined by the user, called agent's confidence level. This factor should be increased with time as the neural network weights settle to the right direction. The outputs are the classified category and/or the words that were considered new or wrong in step2.

- **Step5**) *User-system interaction*: The user response to the output of the previous step is provided as corrected words, a corrected category and/or an acceptance of the agent's decision.

- **Step6**) *On-line learning*: According to the user's feedback, the agent corrects the words, adds new words to the dictionary and updates the neural network weights according to the on-line learning rule.

Table 2 presents some examples of the agent-user interaction.

Table 2. Simulation transcription.

Example (1):

User Input:	I'd like to book an airline ticket to London
Agent Reply:	The input is considered as category 1, travel reservation(C/W?)

New words: No words to correct.

User feedback:	Correct

Example (2):

User Input:	Contact the travel agency to cancel my ticket to London
Agent Reply:	The input is classified into category 3, travel cancellation(C/W?)

New words: agency, contact

User feedback:	Correct

Example (3):

User Input:	Arrange me a trip to Cairo
Agent Reply:	The input can not be identified, unknown(C/W?)

New words: arrange, Cairo

User feedback:	No. Category 1, reservation Cairo
Agent Reply:	Done

6   Summary

We have introduced the task classification unit in an interface agent that relies on the neural network learning techniques to acquire the knowledge it needs to assist its user. It is designed to achieve the properties of interacting with the user through natural language interface and learning over time. The ability to build such systems is based on the assumption of the restricted vocabulary domain. The agent learns gradually to better assist the user. We argued that such a gradual learning approach is beneficial as it allows the user to incrementally become confident with the agent's decision. The implemented task classification unit has been tested with real world data in the travel reservation domain. The results show that the system is able to classify the user's input correctly and learn over time.

References

[1] Maes, P. (1995), "Intelligent software: programs that can act independently will ease the burdens that computers put on people," *Scientific American*, vol. 273, no. 3, pp. 66-68.

[2] Maes, P. (1994), "Agents that reduce work and information overload," *Communications of the ACM*, vol. 37, no. 7, pp. 31-40.

[3] Pun, C. and Li, Y. (1998), "Machine translation with corpus-base support," *Proceedings of Fourth International Conference on Computer Science and Informatics North Carolina*, pp.158-161.

[4] Wooldridge, M. and Jennings, N.R. (1995), "Intelligent agents: theory and practice," *The Knowledge Engineering Review*, vol. 10, no. 2, pp.115-152.

[5] Tsui, K.C., Azvine, B., and Plumbley, M. (1996), "The roles of neural and evolutionary computing in intelligent software systems," *BT Technology Journal*, vol. 14, no. 4, pp. 46-54.

[6] Wasserman, P. (1989), *Neural Computing Theory and Practice*, Van Nostrand Reinhold, New York.

[7] Aboulenien, H.A. and De Wilde, P. (1998), "A simple interface agent," *Proceedings of Fourth International Conference on Computer Science and Informatics North Carolina*, pp. 190-192.

[8] Russell, P. and Norvig, P. (1995), *Artificial Intelligence: A Modern Approach*, 2nd ed., Prentice-Hall, New Jersey.

[9] Zurada, J. (1992), *Introduction to Artificial Neural Systems*, West Publishing Company, St. Paul.

[10] Salton, G. (1989), *Automatic Text Processing*, Addison-Wesley, New York.

CHAPTER 8

RECENT ADVANCES IN NEURAL NETWORK APPLICATIONS IN PROCESS CONTROL

U. Halici[1], **K. Leblebicioglu**[1], **C. Özgen**[1,2], and **S. Tuncay**[1]
Computer Vision and Intelligent Systems Research Laboratory,
[1]Department of Electrical and Electronics Engineering,
[2]Department of Chemical Engineering
Middle East Technical University, 06531, Ankara, Turkey
{halici,kleb,cozgen}@metu.edu.tr, tuncay@ec.eee.metu.edu.tr

> *You must understand the process before you can control it and the simplest control system that will do the job is the best.*
>
> *William L. Luyben (1990)*

This chapter presents some novel applications of neural networks in process control. Four different approaches utilizing neural networks are presented as case studies of nonlinear chemical processes. It is concluded that the hybrid methods utilizing neural networks are very promising for the control of nonlinear and/or Multi-Input Multi-Output systems which can not be controlled successfully by conventional techniques.

1 Introduction

Classical control techniques such as Proportional Integral (PI) control or Proportional Integral Derivative (PID) control are successfully applied to the control of linear processes. Recently, linear Model Predictive Control (MPC) has also successfully been accomplished in the control of linear systems. However, about 90% of the chemical and biological processes are highly nonlinear and most of them are Multi-

Input Multi-Output (MIMO). When the system is nonlinear and/or MIMO the above conventional techniques usually fail to control such systems. Nowadays, the systems used in industry require a high degree of autonomy and these techniques are not capable of achieving this [9].

The need to meet demanding control requirements in increasingly complex dynamical control systems under significant uncertainty makes the use of Neural Networks (NNs) in control systems very attractive. The main reasons behind this are their ability to learn to approximate functions and classify patterns and their potential for massively parallel hardware implementation. In other words, they are able to implement (both in software and hardware) many functions essential to controlling systems with a higher degree of autonomy.

Due to their ability to learn complex nonlinear functional relationships, neural networks (NNs) are utilized in control of nonlinear and/or MIMO processes. During the last decade, application of NNs in identification and control has been increased exponentially [24], [64].

The wide spread of application has been due to the following attractive features:

1. NNs have the ability to approximate arbitrary nonlinear functions;
2. They can be trained easily by using past data records from the system under study;
3. They are readily applicable to multivariable systems;
4. They do not require specification of structural relationship between input and output data.

This chapter contains four different approaches utilizing NNs for the control of nonlinear processes. Each of them is examined as a case study and tested on nonlinear chemical processes. While the first case study is utilizing NN in the usual way, the other three case studies are novel hybrid approaches.

In case study I, a simple NN control system having a neuro-estimator and a neuro-controller is developed to control a neutralization system, which shows a highly nonlinear characteristic. The system is tested for both set point tracking and disturbance rejection. The performance is compared with a conventional PID controller.

In case study II, a new structure, which incorporates NNs with the linear MPC to extend its capacity for adaptive control of nonlinear systems, is proposed. The developed controller is utilized in the control of a high-purity distillation column using an unsteady-state simulator. Its set point tracking and disturbance rejection capabilities are tested and compared with a linear MPC controller.

In case study III, an approach, which incorporates NNs with PI controllers, is presented. The main problem with the PI type controllers is the determination of proportional and integral constants for each operating (bias) point. The neural network is used to make an interpolation among the operating points of the process to be controlled and produce the related integral and proportional constants. The controller is tested in control of a binary batch distillation column.

In case study IV, a new method is proposed to control multi-input multi-output (MIMO) nonlinear systems optimally. An "optimal" rule-base is constructed, which is then learned and interpolated by a NN. This rule-based neuro-optimal controller is tested in the control of a steam-jacketed kettle.

The organization of the rest of this chapter is as follows: in the next two sections the concept of process control and use of NNs in process control are presented. The other next four sections are dedicated to case studies. The last section contains the remarks and future studies.

2   Process Control

In the development, design, and operation of process plants, the process engineers are involved with five basic concepts: state, equilibrium, conservation, rate, and control.

The identification of a system necessitates the definition thermo-dynamic **state** according to which all the properties of a system are fixed. Chemical, physical and biological systems can not be carried beyond the limits of thermodynamic **equilibrium**, which limits the possible ranges of chemical and physical conditions for the processes taking place in the system.

Conservation of mass, momentum and energy require that certain quantities be conserved in the process because of the mass, energy and momentum balances. The type and size specifications of process equipment of a system depend on the amounts of throughput and also on the **rates** at which physical, chemical and biological processes take place in the equipment. This concept is covered in the field of chemical and biological kinetics.

A process can be feasible both thermodynamically and kinetically but can still be inoperable because of poor operating performance. This can be a result of uncontrollability of the process or because of uneconomic conditions. Therefore, **control** of a system for a satisfactory operating performance, physically and economically, is as important for the design and operation of a process system as the concept of equilibrium and rate of processes [25].

Process control is the regulation of chemical, physical and biological processes to suppress the influence of external disturbances, to ensure the stability of the process and to optimize the performance of the process.

Some important features of process control can be listed as [25]:

- The study of process control necessitates first the study of **time-dependent** changes. The problems can not be formulated without a dynamic structure. The control of any process can only be studied by a detailed analysis of the unsteady-state behavior which can be obtained from the dynamic model of the process.

- Also, process control systems are **information-processing** systems. They receive information, digest it, act on it and generate information as signals.

- All process control systems are **integrated** systems of components, in which each component affects the overall performance of the system. Therefore, a global approach which considers the whole system and its environment as an entity is important.

- Most process control systems are **feedback** systems in which information generated by the system are processed again to regulate the behavior of the system.

- Finally, the **economical** concerns should be among the performance objectives of the process control system.

Process control systems in chemical, biological and physical process industries are characterized by constantly changing performance criteria, primarily because of the changes of the market demand. Also, these processes are highly nonlinear and can not be well modeled. Thus, the control has to be done to update the manipulated variables on-line to satisfy the changing performance criteria on the face of changing plant characteristics. Various control techniques based on different performance criteria and process representations are used to solve these problems.

During the operation of a plant, several requirements must be satisfied and can be considered as performance criteria. Some of them are listed below [58]:

1. Safety and environmental regulations,
2. Product specifications,
3. Operational constraints,
4. Economics.

These criteria must be translated to mathematical expressions in order to write a control law. They can further be classified as objectives (functions of variables to be optimized dynamically) and constraints (functions of variables to be kept within bounds).

Translation of performance criteria to mathematical expressions may require some assumptions. These assumptions are made not only to simplify the solution of the problem, but also to make the problem manageable for implementation in the existing hardware.

All controllers use a representation or a model of the process. Generally, in chemical and biological processes, models are nonlinear and also the model parameters are not well known. Thus there is always a mismatch between the model prediction and the actual process output. Additional reasons for the differences are due to changes in operating points and equipment.

Mismatches between a plant and its model result in unsatisfactory trading of the performance criteria. The tuning parameters can help the trade-off between the fast set point tracking and smooth manipulated variable response. It is always desirable to minimize the amount of on-line tuning by using a model of the process at the design stage that includes a description of the uncertainties.

Even if an uncertainty description is used, there is always a need for updating the model parameters on-line in an adaptive way. **Model Predictive Controllers,** MPC, are those controllers in which the control law is based on a process model [17]. MPC is a control scheme in which the controller determines a manipulated variable profile that optimizes some open-loop performance objective on a time interval extending from the current time to the current time plus a prediction horizon [15]. MPC is suitable for problems with a large number of manipulated and controlled variables, constraints imposed on both the manipulated and controlled variables, changing control objectives and/or equipment failure, and time delays. A model of the process is employed directly in the algorithm to predict the future process outputs.

Usually, in many process control problems, system models are not well defined; either they are missing or system parameters may vary with respect to time. NNs are convenient for obtaining input-output models of systems since they are able to mimic the behavior of the system after training them. Even if the NN model or identification may have mismatches with the plant at the beginning, it becomes better and better as the on-line training progresses. Furthermore, on-line training makes the NN model handle the time varying parameter changes in the plant, directly.

By training the NN to learn the "inverse model" of a plant it can be used as a "controller" for the plant. Also, NN controllers can be used in MPC structures both as estimator and/or controller parts.

Since chemical and biological processes are usually very complex, instead of using NN alone in control of these processes, using them together with conventional approaches such as PI or PID control techniques or recent techniques such as rule based expert systems or fuzzy logic, in a hybrid manner, improves the performance of the overall controller.

3   Use of Neural Networks in Control

In control systems applications, feedforward multi-layer NNs with supervised training are the most commonly used. A major property of these networks is that they are able to generate input-output maps that can approximate any function with a desired accuracy. NNs have been used in control systems mainly for system identification and control.

In **system identification**, to model the input-output behavior of a dynamical system, the network is trained using input-output data and network weights are adjusted usually using the backpropagation algorithm. The only assumption is that the nonlinear static map generated by the network can adequately represent the system's dynamical behavior in the ranges of interest for a particular application. NN should be provided information about the history of the system: previous inputs and outputs. How much information is required depends on the desired accuracy and the particular application.

When a multi-layer network is trained as a **controller**, either as an open-loop or closed loop, most of the issues are similar to the identification case. The basic difference is that the desired output of network, that is the appropriate control input to be fed to the plant, is not available but has to be induced from the known desired plant output. In order to achieve this, one uses either approximations based on a mathematical model of the plant (if available), or a NN model of the dynamics of the plant, or, even, of the dynamics of the inverse of the plant. NNs can be combined to both identify and control the plant, thus forming an adaptive control structure.

We will now introduce some basic ways in which NN training data can be obtained in tasks relevant to control [37]:

- **Copying from an existing controller**: If there is a controller capable of controlling a plant, then the information required to train a neural network can be obtained from it. The NN learns to copy the existing controller. One reason for copying an existing controller is that it may be a device that is impractical to use, such as a human expert. In some cases, only some finite input-output command pairs of a desired controller are known. Then an NN can be trained to emulate the desired controller by interpolating these input-output command pairs.

- **System Identification:** In the identification case, training data can be obtained by observing the input-output behavior of a plant. In more complex cases, the input to the model may consist of various delayed values of plant inputs and the network model may be a recursive one.

- **Identification of System Inverse:** In this scheme, input to the network is the output of the plant and the target output of the network is the plant input. Once the plant inverse NN is obtained, it is fed by the desired plant output and its output is then the desired control input to the plant. The major problem with inverse identification is that the plant's inverse is not always well defined.

- **Model Predictive Controller:** First a multi-layer network is trained to identify the plant's forward model, then another NN, i.e., the controller, uses the identifier as the plant's estimator in an MPC structure. This scheme has an advantage of being an adaptive controller, but it necessitates the computation of the Jacobian of the identifier NN.

There are many advanced networks for more complex system identification of control problems. The reader is referred to [3], [4], [37] for a list of references.

The system identification part is the backbone of almost all neurocontroller architectures so we will discuss this concept in more detail for SISO plants suggested in [38]. These models have been chosen for their generality as well as for their analytical tractability. The models of the four classes of plants can be described by the following nonlinear difference equations:

Model I:

$$y_p(k+1) = \sum_{i=0}^{n-1} \alpha_i y_p(k-i) \quad + g(u(k),..,u(k-m+1)) \tag{1}$$

Model II:

$$y_p(k+1) = f(y_p(k),..,y_p(k-n+1)) + \sum_{i=0}^{m-1} \beta_i u(k-i) \qquad (2)$$

Model III:

$$y_p(k+1) = f(y_p(k),..,y_p(k-n+1)) + g(u(k),..,u(k-m+1)) \qquad (3)$$

Model IV:

$$y_p(k+1) = f(y_p(k),..,y_p(k-n+1);u(k),..,u(k-m+1)) \qquad (4)$$

where $(u(k), y_p(k))$ represents the input-output pair of the plant at time k and $f:R^n \to R$, $g:R^m \to R$ are assumed to be differentiable functions of their arguments. It is further assumed that f and g can be approximated to any desired degree of accuracy on compact sets by multilayer NNs. Due to this assumption, any plant can be represented by a generalized NN model.

To identify a plant, an identification model is chosen based on prior information concerning the class to which it belongs. For example assuming that the plant has a structure described by model III, we have two types of identifiers:

1. Parallel model: In this case, the structure of the identifier is identical to that of the plant with f and g replaced by the corresponding NNs, N_1 and N_2 respectively. This model is described by the equation

$$\hat{y}_p(k+1) = N_1(\hat{y}_p(k),..,\hat{y}_p(k-n+1)) + N_2(u(k),..,u(k-m+1)) \qquad (5)$$

2. Serial-parallel model: The model is described by the equation:

$$\hat{y}_p(k+1) = N_1(\hat{y}_p(k),..,\hat{y}_p(k-n+1)) + N_2(u(k),..,u(k-m+1)) \qquad (6)$$

When a plant is identified, a proper controller can be designed based on the identification model. When external disturbances and/or noise are not present in the system, it is reasonable to adjust the control and identification parameters simultaneously. However, when noise and/or disturbances are present, controller parameter updating should be carried out over a slower time scale to ensure robustness.

A number of applications of NNs to process control problems have been reported. A widely studied application involves a nonlinear-model predictive controller [5], [22], [23], [48], [49], [51], [65]. Piovoso *et al.* have compared NN to other modeling approaches for IMC, global linearization and generic model control and they have found that NNs give excellent performance in the case of severe process/model mismatch [46]. Seborg and co-workers have used radial basis function NN for nonlinear control and they have applied their approaches to simulated systems as well as an actual pH process [39], [47], [48], [49], [53]. They have found the NN based controllers to be superior to other methods in terms of their ease of design and their robustness. NNs are often viewed as black box estimators, where there is no attempt to interpret the model structure [61]. NNs have been used in nonlinear process identification [11], in IMC [13], [39], in adaptive control [7], [16], in tuning conventional PID controllers [63], and in both modeling and control of nonlinear systems [16]. The model adaptation of NN based nonlinear MPC has been studied in [29] and [30].

Narendra *et al.* explained how neural networks can be effectively used for identification and control of nonlinear dynamic systems, where an NN is trained by a backpropagation algorithm for adjustment of parameters [38]. They studied multilayer and recurrent NN in a unified configuration for modeling. Simulation studies on low order nonlinear dynamic systems showed that such modeling and control schemes are practically feasible and they proposed that the same methods can also be used successfully for the identification and control of multivariable systems of higher dimensions.

Bhat *et al.* discussed the use of multilayer NN trained by back-propagation algorithm for dynamic modeling and control of chemical processes [6]. They proposed two approaches for modeling and control of nonlinear systems. The first approach utilizes a trained NN model of the system in a model based control work frame and the second approach utilizes an inverse model of the plant extracted using NN in the internal model control structure. They implemented the first approach on a CSTR where pH is the controlled variable. Their results showed that NN is better in representing the nonlinear characteristics of the CSTR than classical convolution model and, also, the controller performance is superior to convolution model-based control.

Willis *et al.* discussed NN models from the process engineering point of view and explained some approaches for use of NN in modeling and control applications [65]. They considered some industrial applications whereby an NN is trained to characterize the behavior of the systems, namely industrial, continuous and fed-batch fermenters, and a commercial scale, industrial, high purity distillation column. They pointed out that NNs exhibit potential as soft sensors. They also explained a methodology for use of NN models in MPC structure to control nonlinear systems. The results of their simulation studies on a highly nonlinear exothermic reactor have indicated that although there are many questions to be answered about NN for optimum utilization (e.g., topology, training strategy, modeling strategy, etc.), NN are a promising and valuable tool for alleviating many current process engineering problems.

Nguyen *et al.* have presented a scheme for use of NNs to solve highly nonlinear control problems. In their scheme, an emulator, which is a multilayered NN, learns to identify the dynamic characteristics of the system [41]. The controller, which is another multi-layered NN, learns the control of the emulator. Then this controller is used in order to control the actual dynamic system. The learning process of the emulator and the controller continues during the control operation so as to improve the controller performance and to make an adaptive control.

Tan described a hybrid control scheme for set point change problems for nonlinear systems [59]. The essence of the scheme is to divide the control into two stages, namely, a coarse control stage and a fine control stage, and use different controllers to accomplish a specific control action at each stage. For the coarse stage, a modified multilayer NN with backpropagation training algorithm is used, which drives the system output into a predefined neighborhood of the set point. The controller then switches to the fine control stage at which a linearization of the system model is identified around the current set point, and is controlled with an appropriated PID controller. Simulation results have shown that there are some difficulties that can be faced in the development of such a hybrid control scheme, such as the criteria for the controller switching stages, and the possibility of abrupt changes in control input in the controller switching phase. The applicability of this control scheme to nonlinear control problems is discussed.

Dreager *et al.* have proposed a new nonlinear MPC algorithm for control of nonlinear systems [13]. For the prediction step, their algorithm utilizes a NN model for a nonlinear plant. They have applied this algorithm to a pH system control and also a level control system. They have compared the performance of their nonlinear MPC algorithm with that of a conventional PI controller on these two systems. Results have indicated that the proposed controller outperforms with respect to the PI controller.

Hamburg *et al.* examined various methods, especially NN, with respect to their use to detect nuclear material diversions, considering speed and accuracy [19]. The NN technique is enhanced with the use of a computer simulation program for creating the training data set. This simulation approach provided the opportunity of including outliers of various types in a data set for training the NN because an actual process data set used for training possibly might not have outliers. They compared the methods on their ability to identify outliers and reduce false alarms. These methods were tested on data sets of nuclear material balances with known removals. The results obtained by the NNs were quite encouraging.

Sablani *et al.* used NNs to predict the overall heat transfer coefficient and the fluid to particle heat transfer coefficient, associated with liquid particle mixtures, in cans subjected to end-over-end rotation [50]. These heat transfer coefficients were also predicted by means of a dimensionless correlation method on the same data set. The results showed that the predictive performance of the NN was far superior to that of dimensionless correlations.

Noriega and Wang presented a direct adaptive NN control strategy for unknown nonlinear systems [43]. They described the system under consideration as an unknown NARMA model, and a feedforward NN was used to learn the system. Taking NN as a neuro model of the system, control signals were directly obtained by minimizing either the instant difference or the cumulative differences between a set point and the output of the neuro model. They applied the method in flow rate control and successful results were obtained.

Since 1990, there are too many academic papers on NN controllers and applications in process control, though there are a few real applications. Nowadays advantages and disadvantages of NNs have been well understood. New studies, such as hybrid structures, are constructed in which NNs can appear in several stages emphasizing their advantages.

In the following sections four case studies are presented to show the applications of NNs in conjunction with other techniques for control of complex processes.

4   Case Study I: pH Control in Neutralization System

pH control problem is very important in many chemical and biological systems and especially in waste treatment plants. The neutralization process is very fast and occurs as a result of a simple reaction. However, from the control point of view it is a very difficult problem to handle because of its high nonlinearity due to the varying gain (in the range of 1 up to 10^6) and varying dynamics with respect to the operating point (see Figure 1). Introduction of NNs in modeling of processes for control purposes is very useful due to their flexibility in applications.

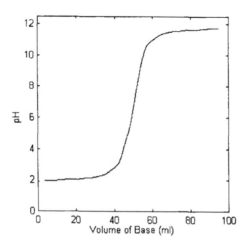

Figure 1. Titration curve of strong acid – strong base system.

In the literature, dynamic mathematical models of pH systems are available [18], [36]. Many control algorithms have been applied to pH control including adaptive, linear model-based, nonlinear internal model, and nonlinear generic model [10], [21], [45], [54], [55].

In this section, a control system having a neuro estimator and a neuro controller is presented and it is used in the control of a pH neutralization system [40].

4.1   Neutralization System

The neutralization system is a highly nonlinear one, whose nonlinearity is reflected in the S shape of the titration curve given in Figure 1. The stirred tank neutralization system that we considered is shown in Figure 2. It has a feed which is composed of one component (acid) and a titrating stream (base). For simplicity, perfect mixing is assumed and the level is kept constant.

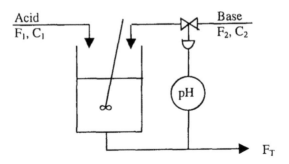

Figure 2. Scheme for the pH process for nonlinear neutralization system.

The material balance can easily be written as [35]

$$V\frac{dC_H}{dt} = F_2 C_2 + F_1 C_1 - (F_1 + F_2) C_H \qquad (7)$$

Assuming the neutralization reaction is very fast, the equilibrium equation can be written as follows [66]:

$$\frac{K_w}{C_H} + C_1 = C_H + C_2 \qquad (8)$$

Using Equations (7) and (8) the change of hydrogen ion concentration can be written as

$$\frac{dC}{dt} = \frac{-C_H^2\left[F_2C_2\text{-}(F_1 + F_2)(\frac{K_w}{C_H}\text{-}C_H)-F_1C_1\right]}{V(K_w + C_H^2)} \tag{9}$$

and

$$pH = -\log(C_H) \tag{10}$$

where

C_1 = concentration of acid (M)
C_2 = concentration of base (M)
F_1 = flow rate of acid (lt/min)
F_2 = flow rate of base (lt/min)
C_H = concentration of hydrogen ion (M)
V = volume of tank (lt)
K_w = water dissociation constant = 1×10^{-14}

Nominal values are
C_{1s} = 0.01 M;
C_{2s} = 0.01 M;
F_{1s} = 0.3 lt/min;
V = 3 lt

C_H is the process state variable while F_2 is selected as the manipulated variable.

4.2 Neural Network Control of the Neutralization System

The structure of the NN controller system is shown in Figure 3. The controller system has a NN controller and a NN estimator.

The estimator is trained by taking the error between the desired plant output and the estimator output. On the other hand the controller is trained by taking the error between the estimator output and a reference point. So the controller assumes that the estimator output matches the plant output.

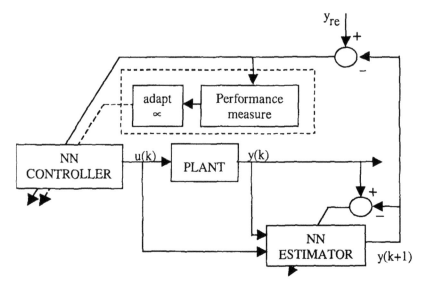

Figure 3. NN controller system used for the neutralization system.

The neural estimator is a multilayer feedforward NN with 10 neurons in the input layer, 20 in the hidden layer and one in the output layer. The values of the initial weights are chosen randomly between –0.1 and 0.1. The backpropagation algorithm is used to train the network. The value of learning rate α is decided by using 1-dimensional search. The input vector for the neuro estimator is chosen as:

$$\mathbf{x}_k = \left[y(k) \; y(k-1)..y(k-m); u(k) \; u(k-1)..u(k-m) \right] \tag{11}$$

and the output is $y_{\text{estimated}}(k+1)$.

After training the neural estimator the controller starts and the window data for the estimator are updated. The neural controller is also a multilayer feedforward NN with 10 neurons in the input layer, 10 in the hidden layer and one in the output layer. The values of the initial weights are chosen randomly between –0.1 and 0.1. The input of the neuro controller is

$$\mathbf{x}_k = \left[y(k) \; y(k-1)..y(k-m); u(k) \; u(k)..u(k-m) \right] \tag{12}$$

and the output is $u(k+1)$.

Again the backpropagation algorithm is used for training the neuro controller; however, learning rate α is chosen as a function of the square of the error, P_k. At sampling time, k, α is calculated as a function of the P_k, according to the set of linguistic rules:

> If P_k is LARGE   then α is 0.1.
> If P_k is MEDIUM   then α is 0.01.
> If P_k is SMALL   then α is 0.001.

The linguistic variables for P_k can be chosen as fuzzy sets, but here they are divided arbitrarily into regions as

> LARGE   = $[25-16]$
> MEDIUM = $[16-09]$
> SMALL   = $[09-00]$

4.3   Results

The NN control system described in Section 4.2 is used for control of the neutralization system. Also a PID controller is designed for comparison. These controllers are compared for set point tracking and disturbance rejection cases.

In set point tracking the initial steady state point in pH is taken as 2.0 and a change of 5.0 is considered to reach a neutral point of pH = 7.0.

In disturbance rejection the system is considered to be at the neutral point at the start as pH of 7.0 and then a −20% load change is given to the flow rate of acid at $t = 25$ min and a +20% change is given to the concentration of the base solution at $t = 100$ min to test the performance of the controllers.

4.3.1   Conventional PID Controller Performance

Tuning of the PID controller is done with Ziegler-Nichols rules [36], [52]. The responses of the system for set point tracking and disturbance rejection are given in Figures 4 and 5. It is seen that the conventional PID controller has failed to control the neutralization system.

4.3.2 NN Controller Performance

The output of the neural estimator in comparison with the actual plant output is shown in Figure 6 for different inputs. The responses of the NN controller for set point tracking and disturbance rejection are given in Figures 7 and 8.

As can be seen in Figure 7 despite the oscillations seen in the first 40 minutes the NNC brings the system to set point and is better than a conventional PID. It is seen from Figure 8 that NNC works better for disturbance rejection compared to set point tracking.

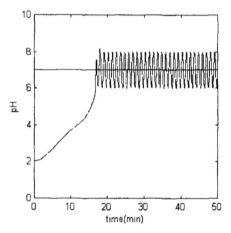

Figure 4. Set point tracking by PID Controller.

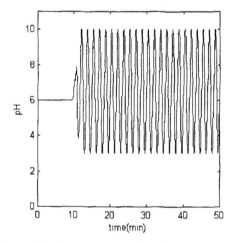

Figure 5. Disturbance rejection by PID Controller.

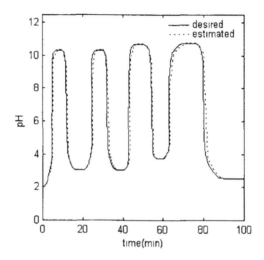

Figure 6. Neural estimator output and actual output for different inputs.

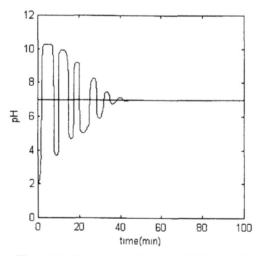

Figure 7. Set point tracking by NN Controller.

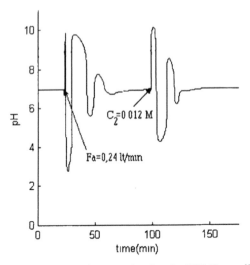

Figure 8. Disturbance rejection by NN Controller.

5 Case Study II: Adaptive Nonlinear-Model Predictive Control Using Neural Networks for Control of High Purity Industrial Distillation Column

In recent years, considerable interest has been devoted to a special class of model based control techniques referred to as Model Predictive Control (MPC) [15], [17]. The basic idea behind MPC algorithm is to use a process model to decide how to adjust the available manipulated variables, in response to disturbances and changing production goals. Control design methods based on the MPC concept have gained high popularity due to their ability to yield high performance control systems. The distinctive feature of the MPC technique is to predict the future behavior of the process outputs based on a non-parametric model, namely, impulse response or discrete convolution model. These can be directly and easily obtained from samples of input-output data without assuming a model structure. Therefore, the MPC technique is especially useful for processes exhibiting unusual dynamic behavior [13].

MPC technique is based on a linear model and, therefore, it is not very well suited for the control of nonlinear systems. Because of this, there have been numerous efforts to extend the linear MPC technique for the control of nonlinear systems [8], [33].

In this work, a new Adaptive Nonlinear-Model Predictive Controller (AN-MPC) utilizing a NN in the MPC work frame is proposed for the adaptive control of nonlinear SISO systems. This technique is used in the control of top-product composition of a distillation column as an application [26], [27].

5.1   Multicomponent High Purity Distillation Column

The performance of the proposed controller is tested on an industrial multi-component high-purity distillation column using an unsteady-state simulation program. The simulation used represents the distillation column in the catalytic alkylation section of the styrene monomer plant of Yarımca Petroleum Refinery, in İzmit, Turkey. In this case study, instead of obtaining the off-line training data from the actual system, the simulator is used because of practical reasons. Since it is a high purity distillation column, it exhibits highly nonlinear characteristics. The unsteady-state simulation package, which is named as DAL, is developed by Alkaya, in 1991 [2].

The distillation column, which has 52 valve trays, was designed to separate Ethyl-Benzene (EB) from a mixture of Ethyl-Benzene, Methyl-Ethyl-Benzene and Di-Ethyl-Benzene having a mole fraction 0.951, 0.012 and 0.037 respectively with a desired top product composition of 0.998. In the process, the top product composition of Ethyl-Benzene is controlled by manipulating the reflux rate as shown in Figure 9.

5.2   Adaptive Nonlinear-Model Predictive Controller Using Neural Networks

5.2.1   Linear Model Predictive Controller

Linear MPC technique may utilize an impulse response model as shown in equation (13) to predict the future behavior of the controlled output as a function of the respective manipulated variable.

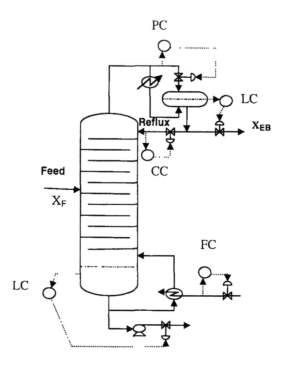

Figure 9. Distillation column.

$$\hat{C}_{n+1} = C_n + \sum_{i=1}^{T} h_i \Delta m_{n+1-i} \qquad (13)$$

where

$\hat{C}_{n+1}$ represents the predicted value of the output for the $n+1^{th}$ sampling,

C_n represents the actual value of the output at n^{th} sampling,

h_i's   represent the impulse response coefficients relating the controlled output to step changes in manipulated variable,

T represents the MPC model horizon, which determines the number of impulse response coefficients,

Δm_i's   represent the implemented step changes in manipulated variable along model horizon prior to $n+1^{th}$ sampling.

Defining r_{n+1} as the set point of the output for the $n+1^{th}$ sampling, the linear MPC law based on equation (13) is formulated as follows:

$$\Delta m_n = (r_{n+1} - C_n - \sum_{i=2}^{T} h_i \Delta m_{n+1-i})/h_i \qquad (14)$$

In equation (14), Δm_n, which is the value of the step change in manipulated variable at n^{th} sampling, is computed to bring the predicted response to set point at $n+1^{th}$ sampling.

5.2.2 Nonlinear-Model Predictive Controller

While the impulse response coefficients, h_i, obtained for a linear system at an operating point can be successfully used for other points, they can only be used for a nonlinear system by local linearization. Thus, there will always be a deviation between the predicted values of the output and the actual system output in nonlinear systems. Therefore, in such systems, this deviation may result in poor control performance when equation (14) is used directly.

However, if the modeling error that comes out at $n+1^{th}$ sampling is estimated somehow, then the linear MPC law can be re-formulated to obtain nonlinear MPC law as given below:

$$\Delta m_n^* = (r_{n+1} - P_{n+1} - C_n - \sum_{i=2}^{T} h_i \Delta m_{n+1-i}^*)/h_i \qquad (15)$$

where Δm^*'s are the step changes in manipulated variable, and P_{n+1} is the deviation at $n+1^{th}$ sampling as defined below:

$$P_{n+1} = C_{n+1} - \hat{C}_{n+1} \qquad (16)$$

5.2.3 Adaptive Nonlinear-Model Predictive Controller via Neural Networks

A Nonlinear-Model Predictive Controller (NMPC) based on equation (15) can be used to control a nonlinear process, unless there is a change in the process conditions. However, if the system parameters change during control operation, then the process model must be adapted to reflect the changes through the use of the estimator for P_{n+1}.

In the NMPC structure, the process is represented by a combination of a linear model and a NN model. The NN used in the NMPC provides an estimate for the deviation between the predicted value of the output computed via linear model and actual nonlinear system output, at a given sampling time. The adaptation of the process model is achieved by updating the NN model via on-line training using the real-time data obtained from the process. Therefore, by continuously training the NN for changes in process dynamics, the NMPC can be used as an Adaptive Nonlinear-Model Predictive Controller (AN-MPC) without any further modification. The resulting AN-MPC structure is shown in Figure 10.

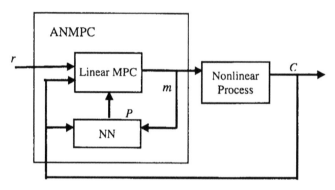

Figure 10. AN-MPC Structure using NN.

In this control structure, the NN is trained at each sampling time, with the present and T previous values of system input-output data with respect to the deviation. Trained NN is used to estimate deviation between the predicted and the actual value of the output. Consequently, AN-MPC computes the value of the manipulated variable, which should be implemented at the present sampling time using equation (15). The NN used in this study is a multi-layer feed-forward NN as shown in Figure 11.

Input vector to NN, $U^k \in R^{2T+1}$, is composed of two sub-input vectors: present and T past values of output and input of the nonlinear system.

$$U^k = \left[C_n, C_{n-1}, ..., C_{n-T}, m_n^*, m_{n-1}^*, ..., m_{n-T}^* \right]^T \tag{17}$$

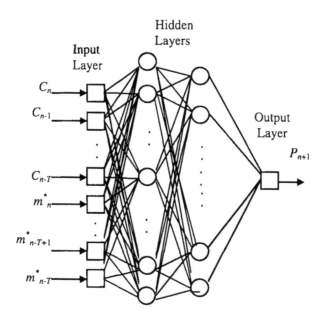

Figure 11. Multilayer NN used in AN-MPC.

As explained before, the NN is trained such that its output vector, which has a single element P_{n+1}, is the deviation of the nonlinear model from its linear MPC model for next sampling.

Since the deviation P_{n+1} is a function of present and past values of the process input and output, these two components of the input vector are shifted in a forward direction at each sampling.

The training of NN is done using backpropagation algorithm. Two types of training strategies, off-line training and on-line training, are used in this particular application. In off-line training the NN is trained to obtain the deviation around an initial operating point prior to control operation. The data required for this are obtained by utilizing a step response experiment where K consecutive step inputs are applied to the system, in the open-loop. That is, the system output resulting from K consecutive step inputs (one step change at each sampling) is observed and compared with the MPC prediction, at each sampling. The difference among them constitutes the deviation for each sampling.

In on-line training NN is continuously trained to obtain the deviation using on-line data for adaptive control purposes using AN-MPC. Thus at each sampling, actual output is observed and compared with its predicted value to compute the deviation. Then, this input-output and deviation data obtained from the system are used to train the NN, at each sampling.

5.3   Identification

The first step in the development phase of the AN-MPC for the distillation column is identification, where impulse response coefficients representing the relationship between reflux rate and EB composition at the top at a sampling period Δt are determined. Consequently, when a unit step input is given to the reflux rate, the top product EB mole fraction changes from 0.9988 to 0.9995 within 7.06 hours which is the response time of the process (Figure 12).

Therefore, settling time of the process is found to be 6.5 hours. Since, the settling time is too large, the model horizon, T, is chosen as 50. From this, sampling period is calculated as 0.13 hours (7.8 minutes) and impulse response coefficients are determined as given in Table 1.

Having determined the impulse response coefficients and MPC model horizon, T, the discrete convolution model (Equation 13) relating top product EB mole fraction to reflux rate is found, where C and Δm stand for top product EB mole fraction and step change in reflux rate, respectively.

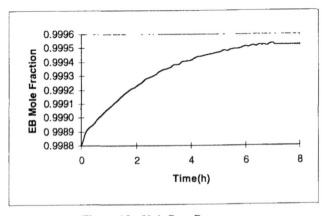

Figure 12. Unit Step Response.

Table 1. Impulse Response Coefficients.

i	$h_I \times 10^5$	i	$h_I \times 10^5$	i	$h_I \times 10^5$	i	$h_I \times 10^5$	i	$h_I \times 10^5$
1	2.7	11	2.8	21	1.2	31	0.6	41	−0.2
2	3.3	12	1.8	22	1.9	32	1.6	42	0.8
3	2.4	13	2.4	23	1.4	33	0.6	43	0.8
4	2.4	14	2.0	24	0.4	34	0.5	44	−0.2
5	3.4	15	1.0	25	1.4	35	0.6	45	0.8
6	2.6	16	2.0	26	1.4	36	0.6	46	0.8
7	2.6	17	1.6	27	0.4	37	0.6	47	−0.2
8	2.6	18	2.2	28	1.4	38	0.5	48	0.8
9	2.3	19	1.2	29	0.8	39	0.8	49	−0.2
10	1.8	20	1.2	30	0.6	40	0.8	50	0.8

5.4   Development of the Neural Network Model

The second step in the development of AN-MPC is the development of a NN model representing the deviation of the linear MPC model from the actual (nonlinear) system through off-line training. This is accomplished in three steps: 1. Obtaining the training data for the NN by utilizing an open-loop step response experiment; 2. Determination of a suitable NN architecture by following a trial and error procedure; 3. Off-line training of NN by using the data obtained in the first step. This enables the NN model to operate satisfactorily at the start. Otherwise, the initial modeling uncertainty for the NN can be too large and the system may become unstable at the beginning of the control operation.

The off-line training data for NN model are obtained through a step response experiment where 50 arbitrary consecutive step changes are introduced to manipulated variables as shown in Table 2, and the response is observed as shown in Figure 13. At each sampling time, by using the linear model of equation (13), the system response, and equation (16), deviation of linear model predictions from the actual output is calculated. Then, using these data the training vectors for the NN are created.

A 10^{-8} order of magnitude error in P_{n+1} results in a 10^{-3} order of magnitude change in the control input, which is acceptable for this application. Therefore training of the NN is terminated when the error in training is less than or equals to 1×10^{-8}. Since the model horizon, T, is chosen as 50, the number of nodes in the input layer is 102. By following a trial-error procedure a suitable NN architecture satisfying the training-stop criteria is determined as a three-layered feed-forward NN, having 104 and 50 nodes in the first and second hidden layers with sigmoid type activation functions, and an output node with an identity type activation function.

Table 2. Step Changes in Reflux Rate.

t (h)	Δm (lbmol/h)	t (h)	Δm (lbmol/h)	t (h)	Δm (lbmol/h)
0.00	2.10	2.21	0.21	4.42	2.10
0.13	2.20	2.34	0.19	4.55	2.20
0.26	−0.20	2.47	0.78	4.68	4.40
0.39	−0.14	2.6	1.78	4.81	−1.10
0.52	−6.00	2.73	1.90	4.94	−2.20
0.65	−4.50	2.86	1.30	5.07	−3.20
0.78	2.10	2.99	1.40	5.20	−0.19
0.91	2.20	3.12	−0.17	5.33	−0.21
1.04	4.40	3.25	−0.19	5.46	−3.45
1.17	−1.10	3.38	−0.70	5.59	−2.78
1.30	−2.20	3.51	−0.50	5.72	−0.19
1.43	−3.20	3.64	2.10	5.85	0.21
1.56	−0.19	3.77	2.20	5.98	0.19
1.69	−0.21	3.90	−0.20	6.11	0.78
1.82	−3.45	4.03	−0.14	6.24	1.80
1.95	−2.78	4.16	−6.00	6.37	1.90
2.08	−0.19	4.29	−4.50		

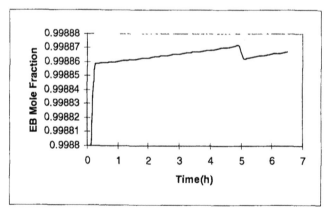

Figure 13. Response of the Distillation Column to changes given in Table 2.

5.5   Control Application

After obtaining the linear MPC model, model horizon and NN model, which represents the deviation of linear MPC model from the actual system, these two models are combined in an MPC workframe. AN-MPC is obtained, in which on-line training of NN is maintained continuously to adapt the controller for changes in process operating conditions. The AN-MPC is tested for its set point tracking and disturbance rejection capabilities. In order to test the performance of the AN-MPC and compare it with that of linear MPC for disturbance rejection capability, the Ethyl-Benzene (EB) mole fraction in feed composition was decreased by 3% (from the steady-state value of 0.9513 to 0.9228), keeping relative mole fractions of Di-Ethyl-Benzene and Methly-Ethyl-Benzene constant. The open-loop response of the process, for this –3% disturbance in the feed composition, is given in Figure 14.

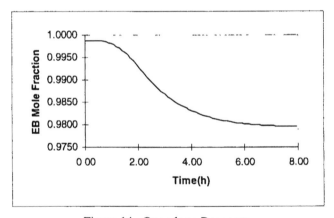

Figure 14. Open-loop Response.

The closed-loop response of the process with the linear MPC and the corresponding control inputs are given in Figures 15 and 16, respectively. The closed-loop response of the process with AN-MPC and corresponding control inputs are given in Figures 17 and 18, respectively.

As it can be seen from Figure 14, when –3% disturbance is introduced to EB mole fraction in feed, the EB mole fraction in top product changes from 0.9988 to 0.9797 within 7 hours. When the linear MPC is used to control the system, the controlled response shows some deviation from set point (Figure 15) and control input is very oscillatory (on-off type) changing between zero reflux and total reflux. Obviously, such behavior of the reflux rate for a distillation column is not practically acceptable. Whereas, when the AN-MPC is used to control the system, the controlled response, as shown in Figure 17, shows little deviation from the set point and, in this case, it matches the set point after 5 hours. Furthermore, the control input (Figure 18) exhibit much smoother behavior than that of linear MPC and they change within reasonable limits.

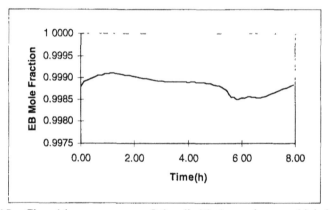

Figure 15. Closed-loop response of the distillation column, which is under control of linear MPC, to a –3% step change in EB Feed Composition.

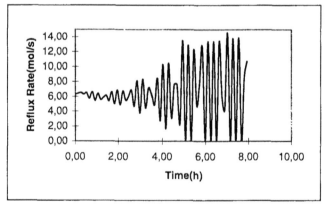

Figure 16. Control actions (reflux rate changes) of linear MPC for a –3% step change in EB feed composition.

In order to test performance of the AN-MPC and to compare it to that of linear MPC for set point tracking capability, the set point is changed in the EB mole fraction from 0.9988 to 0.9900. For this change, the closed-loop response and respective control actions of linear MPC are as shown in Figures 19 and 20, and the closed-loop response and respective control actions with AN-MPC are shown in Figures 21 and 22 respectively.

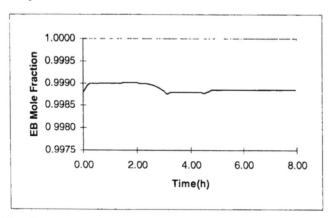

Figure 17.   Closed-loop response of the distillation column, which is under control of AN-MPC, to a –3% step change in EB feed composition.

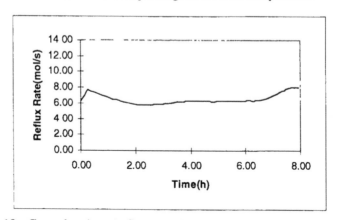

Figure 18.   Control actions (reflux rate changes) of AN-MPC for a –3% step change in EB feed composition mole fraction.

As can be seen from the Figures 19 and 20, the controlled output using linear MPC is oscillatory and does not match with the new set point and the respective control input shows high oscillations. However, control input of AN-MPC is quite smooth and exhibits very small oscillations compared to that of linear MPC. Furthermore, the output controlled with the AN-MPC matches the set point within 6 hours with a very small oscillation compared to that of linear MPC as it can be observed in Figures 21 and 22.

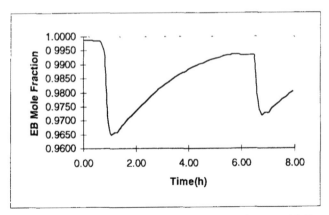

Figure 19. Closed-loop response of the distillation column, with linear MPC, to a set point change of –0.0088 in EB.

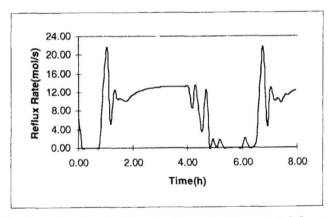

Figure 20. Control actions (reflux rate changes) of linear MPC for a set point change of –0.0088 in EB mole fraction.

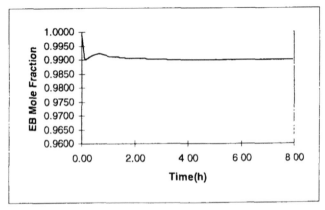

Figure 21. Closed-loop response of the distillation column, with AN-MPC, to a set point change of –0.0088 in EB mole fraction.

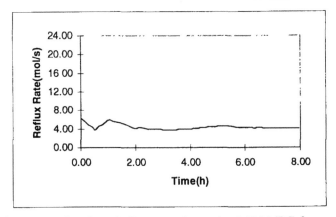

Figure 22. Control actions (reflux rate changes) of AN-MPC for a set point change of –0.0088 in EB mole fraction.

6   Case Study III: PI Controller for a Batch Distillation Column with Neural Network Coefficient Estimator

The main problem with the conventional PI type controllers is the determination of proportional and integral coefficients for each operating (bias) point. In this section, a control method in which a NN is incorporated as an online parameter estimator for the PI-type controller is proposed and used in the control of a binary batch distillation column [60].

6.1   Binary Batch Distillation Column

Batch distillation is an important unit operation where small quantities of high technology/high value added chemicals and bio-chemicals are to be separated. The other separation unit, which is widely used in the chemical industry, is the continuous distillation column. Unlike batch distillation, the mixture, which is separated, is continuously supplied to the column in the continuous distillation case. The most outstanding feature of batch distillation is its flexibility. This flexibility allows one to deal with uncertainties in feed stock or product specification. The operation of a batch distillation column can be described as three periods: start up, production and shutdown periods. The column usually runs under total reflux in the start up period until it reaches the steady

state where the distillate composition reaches the desired product purity
[12], [35].

We will consider a basic separation system as depicted in Figure 23.
This column is used to separate two components in the liquid mixture
by taking advantage of the boiling points; that is, the component with
the lower boiling point will tend to vaporize more readily and therefore
can be selectively collected in the vapor boiled off from the liquid [35].

The basic requirement of the simulation to be developed is to compute
the overhead or distillate composition (condenser product) as a function
of time. If we consider a binary mixture, the lighter component will
have a higher composition in the distillate than in the kettle (bottoms).
However as the total amount of binary is reduced due to continued
withdrawal of the distillate, the concentration in the light component in
the distillate will decrease and get to an eventually low level. This
decrease in the more volatile component concentration while inevitable
can be delayed by increasing the reflux ratio during the distillation at
the expense of the distillate (product) flow rate.

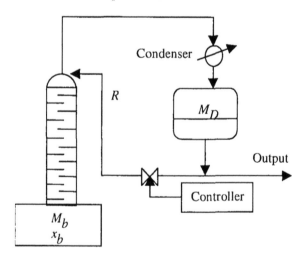

Figure 23. Binary batch distillation column with composition controller.

Dynamic simulation of the batch distillation column and investigation
of an automatic control system for distillate composition have been
done in the study by using the assumptions [35]:

1. Reflux drum and tray holdups are constant;

2. Binary system with constant volatility;
3. Equimolar overflow;
4. Vapor-liquid equilibrium is attained in each tray;
5. Vapor holdup is negligible when compared with liquid holdup.

The variables for the column model are:

$$
\begin{aligned}
H &= \text{Liquid holdup (mole)} \\
G &= \text{Vapor holdup (mole)} \\
L &= \text{Liquid flow rate (mole/sec)} \\
R &= \text{Reflux rate (mole/sec)} \\
M_b &= \text{Kettle holdup (mole)} \\
C &= \text{Condenser holdup (mole)} \\
t &= \text{Time (sec)}
\end{aligned}
$$

At liquid phase, total molar balance for the plate i is given by

$$
\frac{dH_i}{dt} = L_{i+1} - L_i \tag{18}
$$

Since H_i is assumed to be constant, thus $dH_i/dt = 0$, we conclude that $L_{i+1} = L_i = L_{i-1} = L_1 = R$. In nth tray the vapor phase total molar balance gives

$$
\frac{dGi}{dt} = V_{i+1} - V_i \tag{19}
$$

Since the vapor holdup is assumed to be constant, $V_{i+1} = V_i = V_{i-1} = ... = V_1 = V$. Total molar balance for the kettle gives

$$
\frac{dM_b}{dt} = R - V
$$

$$
M_b(0) = M_b{}^0 \tag{20}
$$

Since $dM_b/dt \neq 0$ the total amount of liquid in the kettle changes significantly with time. The component balance for the kettle, trays, and condenser gives:

Kettle:

$$\frac{d(M_b x_b)}{dt} = Rx_1 - Vy_b \tag{21}$$

$$M_b(0) \, x_b(0) = M_b^0 \, x_b^0$$

Plate 1:

$$H \frac{dx_1}{dt} = Rx_2 + Vy_b - Rx_1 - Vy_1 \tag{22}$$

$$x_1(0) = x_b^0$$

where H is the constant liquid holdup on each of the trays. The initial liquid composition on each of the trays is taken as the initial kettle composition x_b^0, which would occur if the column is initially charged with a single liquid.

Plate 2:

$$H \frac{dx_2}{dt} = Rx_3 + Vy_1 - Rx_2 - Vy_2 \tag{23}$$

$$x_2(0) = x_b^0$$

Plate i:

$$H \frac{dx_i}{dt} = Rx_{i+1} + Vy_{i-1} - Rx_i - Vy_i \tag{24}$$

$$x_i(0) = x_b^0$$

Plate N:

$$H \frac{dx_N}{dt} = Rx_d + Vy_{N-1} - Rx_N - Vy_N \tag{25}$$

$$x_N(0) = x_b^0$$

Condenser:

$$C \frac{dx_d}{dt} = Vy_n - Rx_d \tag{26}$$

$$x_d(0) = x_b^0$$

The vapor phase concentrations are calculated from the simple vapor liquid equilibrium relation based on the relative volatility α (Raoult's Law).

$$y_i = \frac{\alpha\, x_i}{1+(\alpha-1)\,x_i} \tag{27}$$

All the equations given above are used to compute the bottoms holdup $M_b(t)$, and still composition $x_b(t)$, the plate compositions $x_1(t), x_2(t), \ldots,$ $x_i(t), \ldots, x_N(t)$ and the distillate composition $x_d(t)$ [35]. In this study, there are 13 trays in the batch distillation column.

There are two basic operation methods for the batch distillation column. The first one is constant reflux rate and variable product composition. The second one is variable reflux and constant product composition of the key component (top product in this case). In this study the aim is to achieve desired constant product composition. Therefore, reflux ratio, R, should be changed during the batch distillation operation. In literature, different methods have been applied to the distillate control problem. The majority of these efforts tried to solve the problem by using optimal control techniques. In these studies, Pontryagin's maximum principle was used in order to maximize the distillate composition [57].

If the column at a bias point is uncontrolled, then the distillate composition $x_d(t)$ would drop off substantially after some time, where it should remain at the relatively high value to give a product of the required purity. In order to remedy this situation we will add to the model the equations for an automatic control system.

Basically a feedback configuration is considered for the control purpose. The term 'feedback' comes from the way in which such a controller works. The variable to be controlled, in this case the distillate composition $x_d(t)$, is sensed (measured) and then compared with the desired value, the set point x_{dset}, to form an error $e(t)$.

$$e(t) = x_{dset} - x_d(t) \tag{28}$$

An ideal controller would keep the error at zero $e(t) = 0$, for which the distillate composition would equal the set point $x_{dset} = x_d(t)$. However, a real controller can not achieve this ideal performance, and it is attempted to design a controller that comes as close as possible to this ideal.

Once the controller generates the error, it is used to modify the manipulated variable within the system to be controlled. In this case the manipulated variable is the reflux rate, R. The manipulation of R will be done according to the controller equation,

$$R = R_{ss} + K_c (e + \frac{1}{T_I} \int_0^t edt) \tag{29}$$

where
   R_{ss} = steady state reflux
   K_c = controller gain
   T_I = controller integral time

This equation describes the action of a proportional integral controller; the first term, $K_c e$, is the proportional part and the second term,

$$\left(K_c / T_I\right) \int_0^t edt \tag{30}$$

is the integral part. We will now consider briefly how each of these sections contribute to the controller of the reflux rate, R. If we had an ideal situation in which the error is always zero, the controller equation simply reduces to $R = R_{ss}$ and the reflux rate would be equal to the steady state value; that is, the batch distillation column would be operating in such a way that the distillate composition is always equal to set point (from error equation).

This situation could never be achieved in practice since the batch distillation column operates in an unsteady state so that $x_d(t)$ is always changing with time and, therefore, does not remain at the set point x_{dset}. If the distillation column has some error $e(t) \neq 0$, the proportional term will change the reflux rate, R, according to the equation $R = R_{ss} + K_c e$. If the error is positive $x_{dset} > x_d(t)$, so that the distillate composition is too low, the proportional control term will increase R, which is the correct action in order to increase $x_d(t)$. On the other hand if the error is negative corresponding to $x_{dset} < x_d(t)$, the reflux rate, R, is reduced by the proportional control term, which again is the correct action to reduce $x_d(t)$. Thus the proportional control action always moves the reflux rate, R, in the right direction to bring $x_d(t)$ closer to the set point x_{dset}. The integral control action removes the offset or steady state error. However, it may lead to an oscillatory response of slowly decreasing amplitude or even increasing amplitude, both of which are undesirable [44]. In the rest of the discussion proportional constant K_c is denoted as K_p and the integral term K_c/T is denoted as K_I. In the following section, the role of the NN in the control method of the batch distillation column is explained. Furthermore, several simulation results are also given.

6.2   PI Controller with Neural Network as a Parameter Estimator

The main problem with the PI type controller is the determination of proportional and integral constants (K_p, K_I) for each operating (bias) point. In order to solve this problem, a NN parameter estimator is incorporated into PI control method as shown in Figure 24.

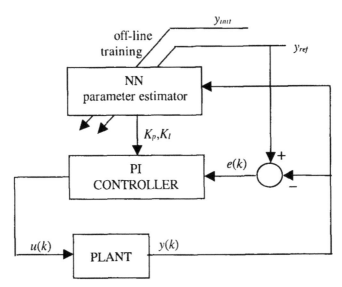

Figure 24. The structure of the PI controller with NN parameter estimator.

The NN shown in Figure 25 is trained for parameter estimation. Actually, the aim of the neural network is to make an interpolation among the operating points of the distillation column and produce the related integral and proportional constants. Hence, a training pair for the neural network is in the form of ($[y_{initial}, y_{ref}]$, $[K_I, K_P]$). The initial and desired bias points actually refer to the initial material concentration and the desired material concentration at the top tray. After training, the neural network can be used as an online parameter estimator for the PI-type controller. As an alternative point of view, the bias points can be seen as the antecedent and the corresponding integral and proportional constants can be seen as the consequent part of an If-Then rule. In this case NN performs an interpolation in the rule space of the system.

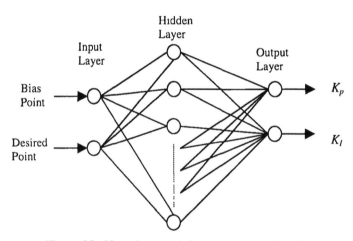

Figure 25. Neural network for parameter estimation.

6.3   Results

In this study, the number of the training pairs is 20 and the training algorithm is the standard backpropagation algorithm. After 50 epochs the mean-square error was reduced to 0.0001. Figure 26 shows a simulation result produced by NN and PI control. The initial concentration for the distillate is 0.5 and the desired concentration is 0.9. NN produced the proportional and integral constants as 8 and 12. It can be seen from the graph that the produced constants yield a satisfactory result. The steady state error is approximately 0.5%.

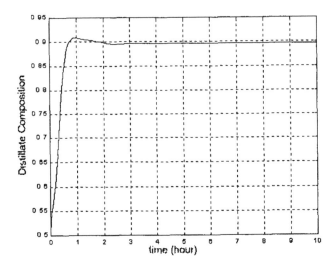

Figure 26. Distillate composition, x_d vs time.

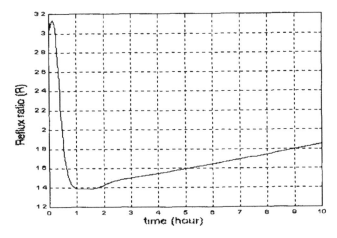

Figure 27. Reflux ratio versus time.

In Figure 27, the corresponding reflux change is given. At the beginning of the operation, reflux ratio decreases in order to increase the distillate composition. After distillate composition reaches its steady state, reflux ratio increases in order to fix the distillate ratio to set point value. It should be noted that, since batch distillation column is used in this study, the material in the still decreases with time. However, the simulation duration is not sufficiently long to observe the fall in the composition in our cases.

Since the amount of maximum overshoot is small, the relative stability of the system is quite good. Figure 28 shows the other two simulations. In the upper part of the Figure 28a, initial distillate composition is 0.5 and the desired (final) composition is 0.8. NN produced the proportional and integral constants as 30 and 20 respectively. In this case, the steady state error is zero; so the PI controller with estimated parameters worked better than the case as shown in Figure 26. The lower part of the Figure 28b shows the simulation results with the initial composition 0.5 and final composition 0.85. It can be seen that desired composition is achieved by the PI control with the help of the NN.

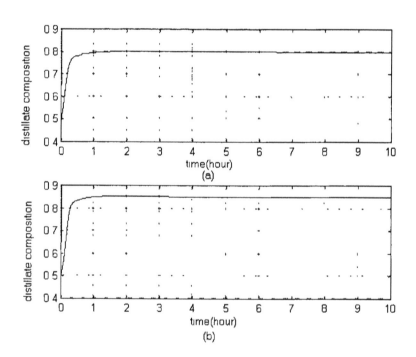

Figure 28. Distillate composition versus time.

In this study, controller parameters are tuned experimentally to achieve fast rise time and small steady-state error and they are used in training the NN. However, some conventional techniques such as the Ziegler-Nichols method can be used for tuning.

7 Case Study IV: A Rule-based Neuro-Optimal Controller for Steam-Jacketed Kettle

In this section, a new method is proposed for the optimal control of multi-input multi-output (MIMO) systems. The method is based on a rule-base derived optimally, which is then interpolated by neural networks.

The design of controllers for MIMO systems has always been a hard problem even for the linear ones [56]. The only prevailing idea used in the control of linear MIMO system is decoupling, if possible at all. During the last decade there have been serious attacks on this problem by methods that are especially constructed to control nonlinear plants, such as neuro-control and sliding mode control techniques [1], [34], [42], [56], [62]. Most of these techniques are quite complicated and possibly work for a particular case only.

The fuzzy control techniques had limited application in MIMO systems control mainly because of the facts that the derivation of rules is not easy (usually not available) and the number of rules is usually large, depending on the number of outputs and states.

Ours is a new attempt to this unsettled problem using a rule-base combined with neural networks. On the other hand there are interesting details and generalizations which will be discussed in the following sections.

7.1 Analysis of the Kettle

The steam-jacketed kettle system has a wide application area in industry. It is especially used in chemical processes. The dynamic response and control of the steam-jacketed kettle shown in Figure 29 are to be considered in this study. The system consists of a kettle through which water flows at a variable rate, w_i kg/min. The inlet water, whose flow rate may vary with time, is at temperature $T_i = 5°C$. The kettle water, which is well agitated, is heated by steam condensing in the jacket at temperature T_V. This is a three-input two-output system.

Flow rate of inlet water, flow rate of outlet water and flow rate of steam are the control inputs of our system. Temperature and the mass of the water inside the kettle are the outputs [12].

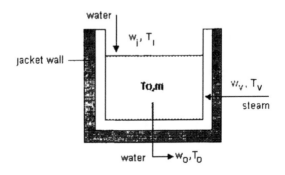

Figure 29. The kettle.

The following assumptions are made for the kettle [12]:

1. The heat loss to the atmosphere is negligible;
2. The thermal capacity of the kettle wall, which separates steam from water, is negligible compared with that of water in the kettle;
3. The thermal capacity of the outer jacket wall, adjacent to the surroundings, is finite, and the temperature of this jacket wall is uniform and equal to the steam temperature at any instant;
4. The kettle water is sufficiently agitated to result in a uniform temperature;
5. Specific internal energy of steam in the jacket, U_v, is assumed to be constant;
6. The flow of heat from the steam to the water in the kettle is described by the expression

$$q = U(T_v - T_o)$$

where
   q = flow rate of heat, J/(min)(m^2)
   U = overall heat transfer coefficient, J/(min)(m^2)(°C)
   T_v = steam temperature, °C
   T_o = water temperature, °C.

The mathematical model of the system can be obtained by making an energy balance on the water side and on the steam side. The symbols used throughout this analysis are defined as follows:

T_I = temperature of inlet water, °C
T_o = temperature of outlet water, °C
w_I = flow rate of inlet water, kg/min
w_o = flow rate of outlet water, kg/min
w_v = flow rate of steam, kg/min
w_c = flow rate of condensate from kettle, kg/min
m = mass of water inside the kettle, kg
m_1 = mass of jacket wall, kg
V = volume of the jacket steam space, m^3
C = heat capacity of water, J/(kg)(°C)
C_1 = heat capacity of metal in jacket wall, J/(kg)(°C)
A = cross sectional area for heat exchange, m^2
t = time, min
H_v = specific enthalpy of steam entering, J/kg
H_c = specific enthalpy of steam leaving, J/kg
U_v = specific internal energy of steam in jacket, J/kg
ρ_v = density of steam in jacket, kg/m^3

Energy balance and mass balance equations for the water and steam side can be written as [12]:

$$mC\frac{dT_o}{dt} = w_iCT_i - w_oCT_o + UA(T_v - T_o) \tag{31}$$

$$\frac{dm}{dt} = w_i - w_o \tag{32}$$

$$m_1C_1\frac{dT_v}{dt} = w_v(H_v - H_c) - (U_v - H_c)V\frac{d\rho_v}{dt} - UA(T_v - T_o) \tag{33}$$

$$V\frac{d\rho_v}{dt} = w_v - w_o \tag{34}$$

As can be seen from the equations, the system is a nonlinear one. The state, input and output vectors are:

$$\mathbf{x} = \begin{bmatrix} T_o \\ m \\ T_v \\ \rho_v \end{bmatrix} \qquad \mathbf{u} = \begin{bmatrix} w_l \\ w_v \\ w_o \end{bmatrix} \qquad \mathbf{y} = \begin{bmatrix} T_o \\ m \end{bmatrix} \tag{35}$$

7.2 A Rule-Based Neuro-Optimal Controller for Nonlinear MIMO Systems

7.2.1 MIMO Systems

It is assumed that a MIMO plant is given with a known mathematical model as shown below

$$\begin{aligned} \dot{\mathbf{x}}(t) &= \mathbf{f}(\mathbf{x}(t), \mathbf{u}(t)) \\ \mathbf{y}(t) &= \mathbf{g}(\mathbf{x}(t)) \end{aligned} \tag{36}$$

where $\mathbf{x}(t)$, $\mathbf{f}(\mathbf{x}(t)$, $\mathbf{u}(t)) \in R^n$, $\mathbf{u}(t) \in R^m$ and $\mathbf{y}(t)$, $\mathbf{g}(\mathbf{x}(t)) \in R^p$. The system output $\mathbf{y}(t)$ is supposed to track a reference signal $\mathbf{y}_d(t) \in R^p$.

7.2.2 Rule Derivation

The controller is developed using a rule-base in which the rules are developed by making use of the mathematical model of the plant in an optimal sense. That is, since a model is available, by partitioning the state-space and the output-space and defining a representative for each partition, one can determine the control signals (i.e., rules) optimally, using a suitably chosen cost function.

Suppose that each component of the state vector has N_i, $i = 1, 2, \ldots, n$ regions and the output vector has O_k, $k = 1, 2, \ldots, p$ components. Then there is a total of $(\prod_{i=1}^{n} N_i)(\prod_{k=1}^{p} O_k)$ rules to be derived. If the system state is initially at the ith partition (the representative of which is $\mathbf{x}_i$) and the system's initial and desired states are at partitions O_v and O_k (their representatives are $\mathbf{y}_v$ and $\mathbf{y}_k$, respectively), the associated rule can be found optimally by solving the optimal control problem of minimizing the cost function in time interval $[0, t_f]$

$$J(\mathbf{u}) = \frac{1}{2}\big(\mathbf{y}(t_f) - \mathbf{y}_k\big)^T \mathbf{H}\big(\mathbf{y}(t_f) - \mathbf{y}_k\big) +$$
$$\frac{1}{2}\int_0^{t_f}\big(\mathbf{y}(t) - \mathbf{y}_d(t)\big)^T \mathbf{Q}\big(\mathbf{y}(t) - \mathbf{y}_d(t)\big)dt + \frac{1}{2}\int_0^{t_f}\mathbf{u}(t)^T \mathbf{R}\,\mathbf{u}(t) \tag{37}$$

subject to the state equation

$$\dot{\mathbf{x}}(t) = \mathbf{f}\big(\mathbf{x}(t), \mathbf{u}(t)\big)$$
$$\mathbf{x}(0) = \mathbf{x}_t \tag{38}$$
$$\mathbf{y}(t) = \mathbf{g}\big(\mathbf{x}(t)\big)$$

Usually **H**, **Q** and **R** are diagonal matrices with suitably chosen diagonal entries. The vector function $\mathbf{y}_d(t)$ can be taken as any smooth function with

$$\mathbf{y}_d(0) = \mathbf{y}_v \ , \ \mathbf{y}_d(t_f) = \mathbf{y}_k$$

$$\dot{\mathbf{y}}_d(0) = \dot{\mathbf{y}}_d(t_f) = 0 \tag{39}$$

Furthermore, the constraints on $\mathbf{u}(t)$, that is, $|u_i(t)| \leq B_i$, $i = 1, 2, \ldots, m$, can easily be incorporated in our steepest descent like optimal control problem solver [20]. What is supposed to be done is implicitly an interpolation in the function space of optimal controls. Here, it is assumed that the mapping between the given initial-final partitions and the associated optimal control functions is continuous. Therefore, if the number of the partitions is sufficiently high, the approximation error in constructing the optimal control function by a semi-infinite neural network, to be explained in the next section, will be quite small.

7.2.3   Neural Network

In order to be able to generate the control inputs so that the system output trajectory follows an optimal path between arbitrarily specified initial and final output states, one has to train a multilayer perceptron-like neural network [20]. This neural network should accept present output $\mathbf{y}(t_o)$ and desired output $\mathbf{y}(t_f)$ as its inputs and should generate the optimal control signal $\mathbf{u}(t)$ to accomplish the task. The structure of the controller utilizing NN is shown in Figure 30.

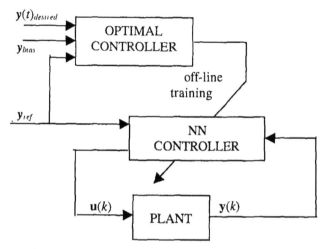

Figure 30. The structure of the rule-based neuro-optimal controller.

For training, input signals produced by optimal control and initial and final points of outputs should be used. It is interesting to note that, at least theoretically, the neural network is a semi-infinite dimensional one [31], [32] in the sense that it is a mapping between the finite dimensional input space and the infinite dimensional output space (i.e., control functions).

The output neurons produce discrete values of input function in $[t_o, t_f]$ interval. Therefore, the neural network can produce the samples of the control signal.

For example, if the number of outputs is n for a single input system, then $\mathbf{y}(t_o)$, $\mathbf{y}(t_f)$ are n-dimensional vectors as

$$\mathbf{y}(t_o)=[\ y_1(t_o)\ \ y_2(t_o)\ ...\ y_n(t_o)]$$

$$\mathbf{y}(t_f)=[\ y_1(t_f)\ \ y_2(t_f)\ ...\ y_n(t_f)]$$

Furthermore, if $[t_o, t_f]$ interval is divided into m parts with sampling period T, a typical training pair is in the form of

$$(\ [\ y_1(t_o)\ y_2(t_o)\ ...\ y_n(t_o)\ ...\ y_1(t_f)\ y_2(t_f)\ ...\ y_n(t_f)\ ],\ [\ u(0)\ u(T)\ ...\ u(mT)])$$

where the $[u(0)\ u(T)\ ...\ u(mT)]$ is the discrete input vector, which moves the system from $\mathbf{y}(t_o)$ to $\mathbf{y}(t_f)$ and is produced by the optimal control. After a training operation, the neural network responds immediately and acts as a real-time controller. In fact, the neural network produces the optimal control vector for the control horizon $[t_{present}, t_{future}]$ at $t_{present}$. The control horizon $t_{future}-t_{present}$ is much larger than the sampling duration. As mentioned already, the mapping between the input-output space and optimal control functions is assumed to be continuous. The data (i.e., the optimal control functions obtained by solving the optimal control problem) represent evaluations of this mapping at particular instants. So, the problem of conflicting rules does not exist.

7.3   Results

In our simulation, the output temperature range is chosen as [5°C, 75°C] and the mass range in the tank is chosen as [10kg, 20kg]. There is no need to partition the rest of the states because these are related with the temperature of the steam entering the jacket. Since the temperature of the steam entering is constant, single partition is enough for these states. We divide temperature range into seven regions and mass range into two regions. Therefore, we get $7\times7\times2\times2 = 196$ rules from optimal control and we use these 196 rules in order to train our neural networks. Since we have three inputs, three neural networks are constructed, each of them has four inputs, two hidden layers having 100 and 50 neurons and an output layer consisting of 25 neurons. The training algorithm is the backpropagation algorithm having a momentum term. After training, neural networks work as a real time controller. For example, with the initial values for outlet water temperature and mass of the water as [10kg, 5°C] and reference inputs as [15kg, 42°C], our results obtained by on-line NN controllers are given in Figures 31-35. For comparison, the results obtained by the optimal control are also shown in these figures. In Figure 36, water temperature in the kettle, which is controlled by a neural network in real time, is given together with the desired trajectory.

According to Figure 36, neuro-controller performance is satisfactory when compared with the optimal controller performance. After the training stage, the neural network can be used as an online controller. In addition, the output of the neural network can be considered as a function, because it estimates the control functions between two sampling (measurement) intervals. Secondly, the control functions are the optimal ones because the training pairs of the neural network consist of control functions produced by solving the associated optimal control problems.

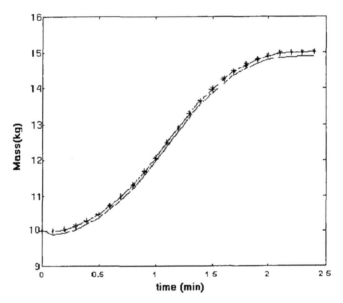

Figure 31. Trajectories for output 1: mass of the water inside the kettle.
* Desired trajectory, - Trajectory from neuro-controller.

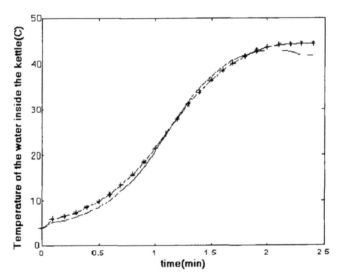

Figure 32. Trajectories for output 2: temperature of the water inside the kettle.
* Desired trajectory, - Trajectory from neuro-controller.

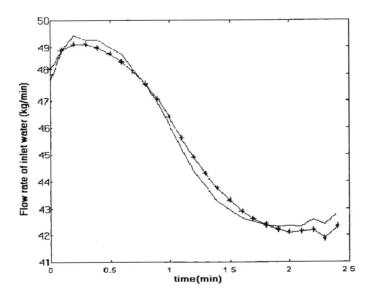

Figure 33. Controlled input 1: flow rate of inlet water.
* Output from optimal control, - Output from neural network.

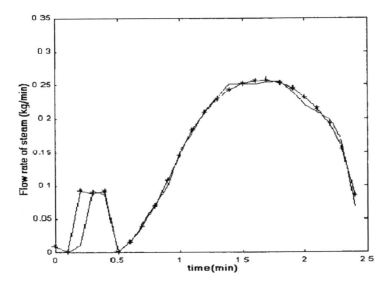

Figure 34. Controlled input 2: flow rate of steam.
* Output from optimal control, - Output from neural network.

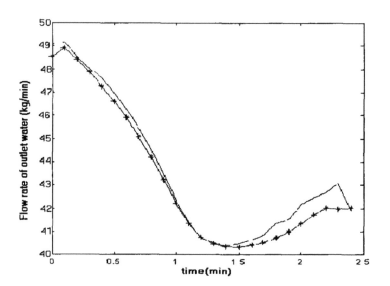

Figure 35. Controlled input 3: flow rate of outlet water.
* Output from optimal control, - Output from neural network.

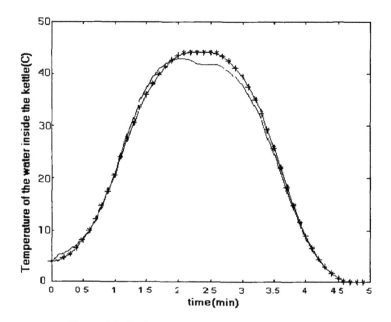

Figure 36. Trajectory for the temperature of water.
* Desired trajectory, - Trajectory produced by on-line NN controller.

8 Remarks and Future Studies

Today's chemical and biological processes in industry are very complex. They are usually nonlinear and/or MIMO. System models of these processes are usually not well defined; either they are missing or system parameters may be time varying. Due to their learning and generalization capabilities, NNs are good candidates for obtaining input-output models of systems. Furthermore, model plant mismatches and the time varying parameter changes in the plant can be overcome by the online training of NNs.

Furthermore, NNs by the "inverse model" of a plant can be used as a "controller" for the plant. Also, NN controllers can be used in MPC structure both as estimator and/or controller parts.

Instead of using NNs alone in control of these processes, they can be combined with conventional approaches such as PI or PID control, optimal control techniques or techniques such as rule based expert systems or fuzzy logic, in a hybrid manner. Such an approach improves the performance of the overall controller.

In this chapter different approaches utilizing neural networks for control of nonlinear processes are presented. Each of them is examined as a case study and tested on nonlinear chemical processes.

In case study I, an NN controller is developed to control a neutralization system which exhibits highly nonlinear dynamics. The controller's performance is tested for both set point tracking and disturbance rejection problems. The NN controller's results are compared with that of the conventional PID controller tuned with Ziegler-Nichols technique. The PID controller failed to control the system by showing oscillatory behavior. However, the NN controller has been able to bring the system to set point, by reducing the oscillations observed at the beginning. Moreover, this NN controller has been able to reject disturbances introduced to the system successfully.

In case study II, linear MPC is used together with NNs to control nonlinear systems. A multilayer NN is used to represent the deviation between the nonlinear system and its linear MPC model. The NN is trained off-line so that the controller operates satisfactorily at the start-up phase. Furthermore, the training of NN is continued on-line using the real-time data obtained from the process. Thus the resultant structure is an adaptive nonlinear MPC controller, AN-MPC.

The performance of the AN-MPC is tested on a simulation of a multi-component high-purity distillation column. Performance tests for disturbance rejection and set point tracking abilities showed that the AN-MPC drives this process quite efficiently, especially in case of set point changes. In contrast, the linear MPC has not been able to control the system for load and set point changes. The success of the hybrid structure, AN-MPC, is because of the fact that the linear MPC determines the coarse control action and NN does the fine tuning. The AN-MPC controller can be generalized further by considering not only the next future sampling instance but next K of them to improve the performance. This generalization is planned as a future work. Current-ly, we are working to extend the structure to control a MIMO plant.

A hybrid control method which is the combination of PI control and NN is introduced in case study III. The method eliminates the controller (PI) tuning problem with the help of the NN. Therefore, it reduces the parameter estimation time at each operating point. The proposed method was tested in the binary batch distillation column and encouraging results were obtained. The hybrid structure of the method uses advantages of each individual method that constructs the hybrid structure. In order to increase the operating range of the proposed controller, the NN must be trained by a large training set which covers the desired wide operating range. However, the major problem is the training pair derivation for NN. In this study training pairs are determined by heuristic methods. For each bias point in the training set, the proportional and integral constants are determined by a trial and error procedure. Therefore, the training pair extraction process can be a time consuming task for engineers who are not experts in batch distillation. Furthermore, the disturbance rejection and robustness issues of the method were not investigated in this study. Hence, they can be studied as a future work.

In case study IV, an optimal neurocontroller has been suggested for controlling MIMO systems. The proposed controller structure was tested by simulation studies on a simple steam-jacketed kettle system.

The preliminary results obtained so far have shown that this method is worth pursuing further. The only disadvantage of the method is that the number of rules to be derived in a complex plant control can be prohibitively large which also makes the derivation time too long. On the other hand, the method is very simple and can be made adaptive with some effort. Studies are continuing to generalize the method to cover the disturbance rejection and robustness problems as well.

All these case studies showed that the hybrid methods utilizing NNs are very promising for the control of nonlinear and/or MIMO systems that can not be controlled by conventional techniques.

Acknowledgments

This work is partially supported under the grant AFP-03-04-DPT-98K12250, Intelligent Control of Chemical Processes.

References

[1] Ahmed, M.S. and Tasaddug, I.A. (1998), "Neural servocontroller for nonlinear MIMO plant," *IEEE Proceedings Control Theory Applications*, vol. 145, pp. 277-291.

[2] Alkaya, D. and Özgen, C. (1991), "Determination of a suitable measurement in an industrial high purity distillation column," *Proc. of AIChE Annual Meeting on Distillation Column Design and Operation.* Los Angeles, CA, pp. 118-124.

[3] Antsaklis, P.J. (1990), "Special issue on neural networks for control systems," *IEEE Control Sys. Mag.*, vol. 10, pp. 3-87.

[4] Antsaklis, P.J. (1992), "Special issue on neural networks for control systems," *IEEE Control Sys. Mag.*, vol. 12, pp. 8-57.

[5] Bhat, N. and McAvoy, T.J. (1989), "Use of neural networks for dynamic modeling and control of chemical process systems," *Proc. Amer. Contr. Conf.*, pp. 1336-1341.

[6] Bhat, N. and McAvoy, T.J. (1990), "Use of neural networks for dynamic modeling and control of chemical process systems," *Comput. and Chem. Eng.*, vol. 14, pp. 573-583.

[7] Boskovic, J.D. and Narendra, K.S. (1995), "Comparison of linear, nonlinear and neural network based adaptive controllers for a class of fed-batch fermentation processes," *Automatica*, vol. 31, pp. 817-840.

[8] Brengel, D.D, and Seider, W.D. (1988), "Multistep nonlinear predictive controller," *Ind. Eng. Chem. Res.*, vol. 28, p. 1812.

[9] Cai, Z.X. (1997), *Intelligent Control: Principles, Techniques and Applications*, World Scientific Publishing.

[10] Chan, H.C. and Yu, C.C. (1995), "Autotuning of gain-scheduled pH control: an experimental study," *Ind. Eng. Chem. Res.*, vol. 34, pp. 1718-1729.

[11] Chen, S., Billings, S.A. and Grant, P.M. (1990), "Nonlinear system identification using neural networks," *Int. J. of Control*, vol. 51, pp. 1191-1199.

[12] Coughanowr, D.R. and Koppel, L.W. (1965), *Process Systems Analysis and Control*, McGraw-Hill.

[13] Cutler, C.R. and Ramaker, B.L. (1979), "Dynamic matrix control – a computer control algorithm," *AIChE National Meeting*, Houston, Texas.

[14] Dreager, A. and Engell, S. (1994), "Nonlinear model predictive control using neural plant models," *NATO-ASI on Model Based Process Control*, Turkey.

[15] Eaton, J.W. and Rawlings, J.B. (1992), "Model predictive control of chemical processes," *Chemical Engineering Science*, vol. 47, pp. 705-720.

[16] Etxebarria, V. (1994), "Adaptive control of discrete systems using neural networks," *IEE Proc. Control Theory Appl.*, vol. 141, pp. 209-215.

[17] Garcia, C.E., Prett, D.M., and Morari, M. (1989) "Model predictive control: theory and practice - a survey," *Automatica*, vol. 25, pp. 335-348.

[18] Gustafsson, T.K., Skrifvars, B.O., Sandstrom, K.V., and Waller K.V. (1995), "Modeling of pH for control," *Ind. Eng. Chem. Res.*, vol. 34, pp. 820-827.

[19] Hamburg, J.H., Booth, D.E., and Weinroth, G.J. (1996), "A neural network approach to the detection of nuclear material losses," *Journal of Chemical Information and Computer Sciences*, vol. 36, pp. 544-553.

[20] Haykin, S. (1996), *A comprehensive Foundation of Neural Networks*, Prentice-Hall, NJ.

[21] Henson, M.A. and Seborg, D.E. (1994), "Adaptive nonlinear control of a pH neutralization process," *IEEE Trans. on Control Systems Technology*, vol. 2, pp. 169-182.

[22] Hernandez, E. and Arkun, Y. (1990), "Neural networks modeling and an extended DMC algorithm to control nonlinear processes," *Proc. Amer. Contr. Conf.*, pp. 2454-2459.

[23] Hernandez, E. and Arkun, Y. (1992), "A study of the control relevant properties of backpropagation neural net models of nonlinear dynamical systems," *Comp. and Chem. Eng.*, vol. 16, pp. 227-240.

[24] Hunt, K.J., Sbarbaro, D., Zbikowski, R., and Gawthrop, P.J. (1992), "Neural networks for control systems – a survey," *Automatica*, vol. 28, pp. 1083-1099.

[25] Johnson, E.F. (1967), *Automatic Process Control*, McGraw-Hill, New York.

[26] Karahan, O., Ozgen, C., Halici, U., and Leblebicioglu, K. (1997), "A nonlinear model predictive controller using neural networks," *Proc. of IEEE International Conference on Neural Networks*, Houston, USA, pp. 690-693.

[27] Karahan, O. (1997), *An Adaptive Nonlinear Model Predictive Controller Using A Neural Network*, M.Sc. Thesis, Chem. Eng. Dept., METU, Ankara, Turkey.

[28] Kirk, D.E. (1970), *Optimal Control*, Prentice-Hall, NJ.

[29] Koulouris, A. (1995), *Multiresolution Learning in Nonlinear Dynamic Process Modeling and Control*, Ph.D. Thesis, MIT.

[30] Koulouris, A. and Stephanopoulos, G. (1997), "Stability of NN-Based MPC in the presence of unbounded model uncertainty," *AIChE Symp. Series*, vol. 93, pp. 339-343.

[31] Kuzuoglu, M. and Leblebicioglu, K. (1996), "Infinite dimensional multilayer perceptions," *IEEE Trans. Neural Networks*, vol. 7, pp. 889-896.

[32] Leblebicioglu, K. and Halici, U. (1997), "Infinite dimensional radial basis function neural networks for nonlinear transformations on function spaces," *Nonlinear Analysis*, vol. 30, pp. 1649-1654.

[33] Li, S., Lim, K.Y., and Fisher, D.G. (1989), "A state-space formulation for model predictive control," *AIChE Journal*, vol. 35, pp. 241-249.

[34] Linkens, D.A. and Nyogesu, H.O. (1996), "A hierarchical multivariable fuzzy controller for learning with genetic algorithms," *Int. Journal of Control*, vol. 63, pp. 865-883.

[35] Luyben, W.L. (1990), *Process Modeling, Simulation and Control for Chemical Engineers*, McGraw-Hill.

[36] McAvoy, T.J. (1972), "Time optimal and Z-N control," *Ind. Eng. Chem. Process Res.*, vol. 11, pp. 71-78.

[37] Miller, W.T., Sutton, R.S., and Werbos, P.J. (1990), *Neural Network for Control*, MIT Press.

[38] Narendra, K.S. and Parthasarathy, K. (1990), "Identification and control of dynamical systems using neural networks," *IEEE Trans. on Neural Networks*, vol. 1, pp. 1-16.

[39] Nehas, E.P., Henson, M.A., and Seborg, D.E. (1992), "Nonlinear internal model control strategy for neural network models," *Comput. and Chem. Eng.*, vol. 16, pp. 1039-1057.

[40] Nesrallah, K. (1998), *Simulation Study of an Adaptive Fuzzy Knowledge Based Controller and Neural Network Controller on a pH System*, M.Sc. Thesis, Chem. Eng. Dept., METU, Ankara, Turkey.

[41] Nguyen, D.H. and Widrow, B. (1991), "Neural networks for self-training control system," *Int. J. Control*, vol. 54, pp. 1439-1451.

[42] Nie, J. (1997), "Fuzzy control of multivariable nonlinear servo-mechanisms with explicit decoupling scheme," *IEEE Trans. on Fuzzy Systems*, vol. 5, pp. 304-311.

[43] Noriega, J.R. and Wang, H. (1998), "A direct adaptive neural network control for unknown nonlinear systems and its application," *IEEE Trans. on Neural Networks*, vol. 9, pp. 27-34.

[44] Ogata, K. (1990), *Modern Control Engineering*, Prentice-Hall Inc.

[45] Palancar, M.G., Aragon, J.M., Miguens, J.A., and Torrecilla, J.S. (1996), "Application of a model reference adaptive control system to pH control: effects of lag and delay time," *Ind. Eng. Chem. Res.*, vol. 35, pp. 4100-4110.

[46] Piovoso, M., Kosanovich, K., Rohhlenko, V., and Guez, A. (1992), "A comparison of three nonlinear controller designs applied to a nonadiabatic first-order exothermic reaction in a CSTR," *Proc. Amer. Cont. Conf.*, pp. 490-494.

[47] Pottman, M. and Seborg, D. (1993), "A radial basis function control strategy and its application to a pH neutralization process," *Proc. 2nd European Cont. Conf.*, pp. 206-212.

[48] Pottman, M. and Seborg, D. (1992), "Identification of nonlinear processes using reciprocal multiquadratic functions," *J. Process Control*, vol. 2, pp. 189-203.

[49] Pottmann, M. and Seborg, D. (1992), "A nonlinear predictive control strategy based on radial basis function networks," *Proc. IFAC DYCORD Symposium*, pp. 536-544.

[50] Sablani, S.S., Ramaswamy, H.S., Sreekanth, S., and Prasher, S.O. (1997), "Neural network modeling of heat transfer to liquid particle mixtures in cans subjected to end-over-end processing," *Food Research International*, vol. 30, pp. 105-116.

[51] Saint-Donat, J., Bhat, N., and McAvoy, T.J. (1991), "Neural net based model predictive control," *Int. J. Control*, vol. 54, pp. 1453-1468.

[52] Seborg, D.E., Edgar, T.F., and Mellichamp, D.A. (1989), *Process Dynamics and Control*, Wiley Series, NY.

[53] Seborg, D. (1994), "Experience with nonlinear control and identification strategies," *Proc. Control'94*, pp. 217-225.

[54] Sing, C.H. and Postlethwaite, B. (1997), "pH control: handling nonlinearity and deadtime with fuzzy relational model-based control," *IEE Proc. Control Theory Appl.*, vol. 144, pp. 263-268.

[55] Skogestad, S. (1996), "A procedure for SISO controllability analysis – with application to design of pH neutralization processes," *Computers Chem. Eng.*, vol. 20, pp. 373-386.

[56] Skogestad, S. and Postlethwaite, I. (1997), *Multivariable Feedback Control*, John Wiley & Sons.

[57] Sorensen, E. and Skogestad, S. (1996) "Optimal startup procedures for batch distillation," *Computers and Chem. Eng.*, vol. 20, pp. 1257-1262.

[58] Stephanopoulos, G. (1984), *Chemical Process Control: An Introduction to Theory and Practice*, Prentice-Hall Int., NJ.

[59] Tan, S. (1992), "A combined PID and neural control scheme for nonlinear dynamical systems," *Proc. SICICI 92*, pp. 137-143.

[60] Tuncay, S. (1999), *Hybrid Methods in Intelligent Control*, M.Sc. Thesis, EE Eng. Dept., METU, Ankara, Turkey.

[61] Ungar, L.H., Hartman, J.E., Keeler, J.D., and Martin, G.D. (1996), "Process modeling and control using neural networks," *Int. Conf. on Intelligent Systems in Process Engineering, AIChE Symp. Series*, pp. 312-318.

[62] Utkin, V.I. (1992), *Sliding Modes in Control Optimization*, Springer-Verlag, Berlin.

[63] Wang, M. and Li, B.H. (1992), "Design of a neural network based controller for control system," *Proc. of SICICI'92*, pp. 1333-1339.

[64] Widrow, B. and Lehr, M.A. (1990), "30 years of adaptive neural networks: perception, madaline and backpropagation," *Proceedings of the IEEE*, vol. 78, pp. 1441-1457.

[65] Willis, M.J., DiMassimo, C., Montague, G.A., Tham, M.T., and Morris, A.J. (1991), "Artificial neural networks in process engineering," *IEE Proc-D*, vol. 138, pp. 3-11.

[66] Wright, A.W. and Kravaris, C. (1991), "Nonlinear control of pH processes using the strong acid equivalent," *Ind. Eng. Chem. Res.*, vol. 30, pp.1561-1572.

CHAPTER 9

MONITORING INTERNAL COMBUSTION ENGINES BY NEURAL NETWORK BASED VIRTUAL SENSING

R.J. Howlett, M.M. de Zoysa, and **S.D. Walters**
Transfrontier Centre for Automotive Research (TCAR)
Engineering Research Centre, University of Brighton
Brighton, U.K.
R.J.Howlett@Brighton.ac.uk

Over the past two decades the manufacturers of internal-combustion engines that are used in motor vehicles have been very successful in reducing the harmful side effects of their products on the environment. However, they are under ever-increasing pressure to achieve further reductions in the quantities of polluting gases emitted by the engine, and a decrease in the amount of fuel consumed per kilometer. At the same time, vehicle characteristics that are desirable to the driver must not be compromised. Satisfying these diverse requirements requires precise engine control and comprehensive monitoring of the operational parameters of the power unit. Engines are highly price sensitive, and it is desirable to achieve the increased level of measurement that is required for enhanced control without additional sensory devices. Thus, the indirect estimation of quantities of interest using virtual-sensor techniques, without direct measurement using dedicated sensors, is a research area with considerable potential. Intelligent-systems techniques, such as neural networks, are attractive for application in this area because of their capabilities in pattern recognition, signal analysis and interpretation. For this reason, the use of neural networks in the monitoring and control of motor vehicle engines is an area of research which is receiving increasing attention from both the academic and commercial research communities. A virtual-sensor technique, the Virtual Lambda Sensor, is described here which uses a neural network for the estimation of air-fuel ratio in the engine.

1   Introduction

The internal-combustion engine is likely to be the most common motor-vehicle power plant until well into the twenty-first century, although new variants such as the Gasoline Direct Injection (GDI) and High Speed Direct Injection (HSDI) Diesel engines may supplant more conventional engine variants.

There are two recurrent themes in the area of automotive engine design: fuel economy and the reduction of harmful emissions from the exhaust. The emission of exhaust gases from Internal-Combustion (IC) engines is a major cause of environmental pollution. In addition the exhaust contains carbon dioxide, which is believed to contribute to the greenhouse effect and global warming. To reduce damage to the environment, governments in the United States, Europe, and parts of the rest of the world have introduced regulations that govern the permissible levels of pollutant gases in the exhaust. All manufacturers of motor-vehicles are required to undertake measures to ensure that their vehicles meet emission standards when they are new. In addition, the vehicle owner is required to ensure that the vehicle continues to meet in-service standards, by submitting it to periodic testing during routine maintenance. In the future, an on-board diagnostic system must be provided which carries out continuous monitoring.

Emission standards have been tightened progressively for over twenty years, to the point where emissions have been reduced by approximately an order of magnitude, measured on a per-vehicle basis. However, regulations are becoming even more stringent. Although existing methods of emission control are adequate to meet current regulations, they need improvements to enable them to meet future legislation [1]. The 1998 California Clean Air Act requires 10% of a manufacturer's fleet to be zero-emission vehicles (ZEVs), an 84% decrease in hydro-carbon (HC) emissions, a 64% decrease in oxides of nitrogen (NOx) output and a 60% reduction in carbon monoxide (CO) production for the entire fleet by the year 2003 [2].

On January 1, 1993, mandatory emission standards were introduced in Europe. This required all new petrol (gasoline) fueled vehicles in

Europe to be fitted with three-way auto-catalysts, thus bringing European standards to comparable levels with the US standards that had been introduced in the 1980s. In 1997, the second stage of regulations was brought into effect which covered both petrol and diesel vehicles. These regulations brought European standards into conformance with US standards up to 1996. The third stage of regulations, which sets standards for year 2000 and beyond, has been proposed. These regulations, when brought into effect, will require petrol-fueled vehicles with electronically controlled catalytic converters to be fitted with on-board diagnostic systems [1].

2   The Engine Management System

In order to achieve these standards it is necessary to maintain strict control of the operating parameters of the engine using a microprocessor-based Engine Management System (EMS) or Engine Control Unit (ECU). The EMS implements control strategies which aim to achieve optimum efficiency and high output power when required, while at the same time maintaining low emission levels. At the same time, in a spark-ignition engine, the EMS must operate the engine in a region favorable to the operation of a three-way catalytic converter, which further reduces the harmful content of the exhaust. The engine must also exhibit good transient response and other characteristics desirable to the operator, known among motor manufacturers as *driveability,* in response to movements of the driver's main control, the throttle or accelerator pedal. The EMS governs the amount of fuel admitted to the engine (via the fuel-pulse width), the point in the engine-cycle at which the mixture is ignited (the ignition timing), the amount of exhaust gas recirculated (EGR), and other parameters in advanced engine designs, for example, the valve timings. It determines values for these parameters from measured quantities such as speed, load torque, air mass flow rate, inlet-manifold pressure, temperatures at various points, and throttle-angle. Figure 1 illustrates the function of the EMS, which must essentially determine values for the *Controlled Variables* from a knowledge of *the Measured Variables*, in order to achieve the *System Aims*.

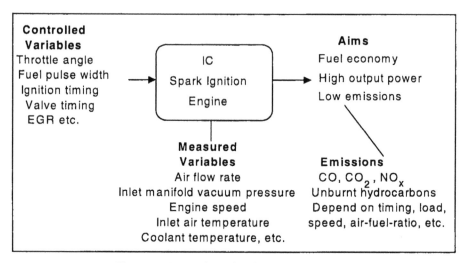

Figure 1. Internal combustion engine control.

The exact detail of the strategies which are used in commercial EMS products is a secret which is guarded closely by the manufacturers. One method which can be used for the selection of fuel pulse width and ignition timing values involves the use of *maps* which are look-up tables held in ROM. The EMS measures the engine speed using a sensor on the crankshaft and estimates the load, often indirectly from the inlet manifold (vacuum) pressure. These values are then used as indices for the look-up tables. Algorithmic and mathematical methods are also used. Research is taking place to develop improved engine control by incorporating neural networks and other intelligent-systems techniques into the EMS.

It has been mandatory in the US for some time, and now is also required in Europe, that, in addition to engine control, the EMS is required to perform *on-board diagnostic* (OBD) functions. Future OBD systems will be required to warn the driver, by means of a *malfunction indicator lamp* (MIL), of faults in the emission-control system which could lead to emission levels that are greater than those permitted.

The high level of accurate control necessary for engines to meet emissions standards requires that the EMS is supplied with comprehensive information about the operational parameters of the engine. Modern engines are equipped with a range of sensory devices which enables the measurement of quantities of interest. Speed, manifold pressure, air mass flow rate, temperature at various points such as the air inlet are examples of quantities that are measured in many engines. In addition, parameters such as crank-angle and camshaft position are measured on more sophisticated power units. Accurate measurement of the ratio between the masses of injected petrol and air, known as the *air-fuel ratio*, is very valuable as an indicator of the point on its characteristics at which the engine is operating. Accurate air-fuel ratio measurement is difficult to achieve economically using conventional methods.

3   Virtual Sensor Systems

As engine control increases in sophistication the number of engine parameters which must be measured also increases. However, manufacturers are reluctant to install new sensors in the engine because of economic considerations. Engines are extremely price sensitive and additional sensors can only be economically justified if they provide very considerable improvements which could not be otherwise attained. Techniques which allow deductions to be made about quantities of interest without the installation of new sensors, by interpreting data from existing sensory devices in a new way, are especially valuable in this respect. The *virtual-sensor* technique allows an estimate to be made of a quantity of interest without the necessity for a sensor dedicated to the measurement. An example, which is described later in this chapter, is the Spark Voltage Characterization method of estimating the air-fuel ratio in the engine cylinder by analysis of the voltage signal at the spark plug.

Virtual-sensor systems require abilities in the domains of pattern-recognition, signal analysis and modeling. Neural networks have been shown to possess distinct strengths in these areas. For example, a neural network based virtual-sensor system is described in the literature that allows the prediction of emission levels from commonly measured quantities [3].

4   Air-Fuel Ratio

A parameter that is of considerable importance in determining the operating point of the engine, its output power and emission levels is the air-fuel ratio (AFR). The air-fuel ratio is often defined in terms of the *excess air factor*, or *lambda* ratio:

$$\text{lambda} = \text{AFR} / \text{AFR}_{st} \tag{1}$$

where AFR = the current air-fuel ratio
and AFR_{st} = the *stoichiometric* air-fuel ratio

Lambda is defined such that a lambda-ratio of unity corresponds to an air-fuel ratio of approximately 14.7:1 at normal temperature and pressure, when the fuel is petrol or gasoline. This is termed the *stoichiometric ratio*, and corresponds to the proportions of air and fuel which are required for complete combustion. A greater proportion of fuel gives a lambda-ratio of less than unity, termed a *rich* mixture, while a greater proportion of air gives a lambda-ratio of greater than unity, termed a *weak* or *lean* mixture. Maximum power is obtained when the lambda-ratio is approximately 0.9 and minimum fuel consumption occurs when the lambda-ratio is approximately 1.1.

Current engines reduce emission levels to within legislative limits by converting the exhaust gases into less toxic products using three-way catalytic converters. For optimum effect, three-way catalytic converters require that the lambda-ratio is closely maintained at the stoichiometric ratio (unity). In modern engines, a *lambda-sensor*, mounted in the exhaust stream, determines whether the lambda is above or below unity from the amount of oxygen present. The EMS uses this to adjust the fuel pulse width to keep the lambda-ratio approximately at unity. Power units currently under development, for example the gasoline direct injection (GDI) engine, may involve operation in lean-of-stoichiometric regions of the characteristics of the engine. Precise control of the air-fuel ratio is of considerable importance here also [4].

The lambda-sensor that is installed in most production vehicles has a voltage-lambda characteristic which effectively makes it a binary device. It can be used to indicate whether the value of lambda is above or below unity, but it is unable to provide an accurate analogue measurement of air-fuel ratio. Accurate measurements can be made using what are referred to as *wideband* lambda-sensors, but they are very expensive, and in fact, even the currently used binary lambda-sensor represents an undesirable cost penalty.

The Spark Voltage Characterization method, described in detail later in this Chapter, allows the air-fuel ratio to be estimated from an analysis of the voltage signal at the spark plug, and so potentially offers the advantage that it permits the elimination of the lambda-sensor.

5   Combustion Monitoring Using the Spark Plug

Although it is not usually considered as a sensor, the spark plug is in direct contact with the combustion processes which are occurring in the cylinder. The use of the spark plug as a combustion sensor in spark ignition (SI) engines offers a number of advantages over other sensory methods. Many comparable techniques, such as pressure measurements or light emission recording by fiber-optics, require that the combustion chamber is modified; this can itself affect the combustion processes. Secondly, the price sensitivity of engines demands that the installation of a new sensor must result in very considerable improvements for it to be economically justifiable. The spark plug is already present in a spark ignition engine, eliminating the need to make any potentially detrimental modifications to the cylinder head, or combustion chamber, and avoiding additional costs which would result from the installation of new equipment. As the spark plug is in direct contact with the combustion, it is potentially an excellent observer of the combustion process. Analyzing the spark plug voltage (and possibly current) waveforms, therefore, potentially provides a robust and low-cost method for monitoring phenomena in the combustion chamber.

A method of using the spark plug as a combustion sensor which has received attention in the literature is known as the *Ionic-Current* method. This has been investigated for measuring combustion pressure, AFR and for the detection of fault conditions such as misfire and knocking combustion. In the ionic-current system, the spark plug is used as a sensor during the non-firing part of the cycle. This is done by connecting a small bias voltage of about 100 volts to the spark plug and measuring the current. This current is due to the reactive ions in the flame which conduct current across the gap when the voltage is applied. The ions are formed during and after combustion, and the type and quantity of ions present depend on the combustion characteristics. The ionization current is also dependent on the pressure, temperature, etc. and therefore is rich in information but very complex [5]. Much work has been done on the use of ionic-currents for monitoring combustion, mainly to estimate combustion pressure, and so the method can act as a replacement for combustion-pressure sensors. Ionic-current systems have also been proposed for AFR and ignition-timing estimation, and misfire and knocking detection [6], [7]. More recently, neural networks have been applied to the analysis of ionic-current data for spark-advance control and AFR estimation [8], [9].

The ionic-current method appears attractive because only minor modifications are required to adapt the engine. However, high-voltage diodes or other switching methods are needed to isolate the ionic-current circuitry from the ignition system, when the high voltage is generated to initiate combustion. These have been prone to failure in the past. The 100V power supply is also an additional component which is required at additional expense.

A second spark plug based sensor technique, which is covered in depth in this chapter, is termed *Spark Voltage Characterization* (SVC). The SVC technique has a number of features in common with the ionic-current method. The SVC method involves the analysis of the time-varying voltage that appears across the spark plug, due to the ignition system, for monitoring combustion phenomena in the cylinder. This analysis can be carried out using a neural network. Using the spark plug as the combustion sensor, this technique has many of the advantages of the ionic-current method. However, as the SVC method involves analyzing the ignition voltage waveform itself, it eliminates the need for an additional bias power supply, and for the associated high-voltage switching circuitry. The use of SVC for estimating the in-cylinder air-fuel ratio is described later in this chapter.

6   The Ignition System of a Spark-Ignition Engine

Figure 2 shows the essential elements of an inductive-discharge ignition-system, as typically installed in a spark-ignition engine. The ignition-coil is essentially a high-voltage transformer, increasing the battery voltage (approximately 12V) to an extra high tension (EHT) pulse. This high voltage creates a spark between the contacts of the spark plug and initiates combustion. The contact-breaker was once a mechanical component in almost all engines, but in modern electronic ignition systems, it is replaced by a semiconductor switch such as an automotive specification transistor or thyristor.

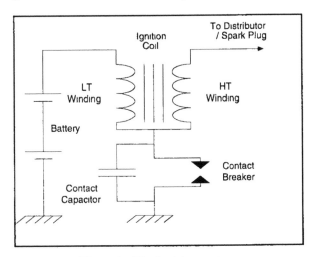

Figure 2. The ignition system.

The contact-breaker closes and current builds up in the low-tension (LT) winding of the coil resulting in the storage of energy; however, the speed at which this occurs is limited by the resistance of the coil. At an appropriate point in the engine-cycle, when an air-fuel mixture has been injected into the cylinder via the inlet-valve (in a port injection engine), and compressed so that the piston lies just before top-dead-center, the contact-breaker opens. The magnetic field in the coil collapses rapidly, with an equally rapid change in magnetic flux, and a high-voltage pulse is induced into the high-tension (HT) winding of the coil. A pulse of approximately 10kV appears across the spark plug terminals, igniting the petrol-air mixture. The resulting combustion drives the power stroke of the engine.

Each cylinder in a four-stroke engine experiences one power stroke for every two revolutions of the crankshaft. In a multi-cylinder engine a mechanical switch geared to the crankshaft and known as a distributor is often used to switch the ignition-pulse to the correct cylinder. Alternative systems make use of multiple coils instead of a distributor. In a *dual-spark* or *wasted-spark* system each cylinder receives a spark once every crankshaft revolution instead of every 720 degrees of rotation. This requires multiple coils, in a multi-cylinder engine, but enables the distributor to be eliminated, and is common practice. Single-cylinder engines also commonly use this principle, as it allows the ignition system to be triggered directly from the crankshaft.

Figure 3 illustrates the spark-voltage waveform obtained from a typical ignition system. The spark plug voltage waveform has a number of predictable phases. As the EHT pulse is generated by the ignition-system the potential difference across the gap rises to between approximately six and 22 kV, before breakdown occurs. Breakdown is accompanied by a fall in voltage, giving a characteristic voltage spike of approximately 10 μs in duration. This is followed by a glow-discharge region of a few milliseconds duration, which appears as the tail of the waveform.

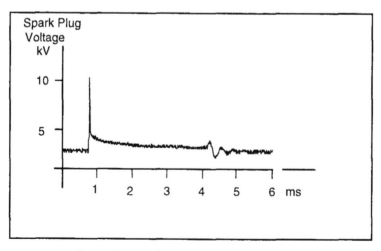

Figure 3. A typical spark voltage waveform.

Empirical observation of the spark plug voltage characteristic has shown that variations in engine parameters lead to changes in the shape of the voltage characteristic. It is predictable that the time-varying voltage exhibits certain major features, for example, a large peak early in the waveform. However, it is not easy to predict the detailed variations that occur as the engine parameters are varied. The signal-to-noise ratio is poor and random variations occur between sparks even when the operating parameters of the engine are kept constant.

The breakdown voltage across the electrode-gap of a spark plug in an operating IC engine is dependent on the interactions of many parameters, for example, the combustion chamber and electrode temperatures, the compression pressure, the electrode material and configuration, and the composition of the air-fuel gas mixture [10], [11]. All of these factors may be attributed to physical properties and processes; for example, the composition of the air-fuel mixture influences the breakdown voltage mainly through temperature and pressure changes.

The spark plug cathode electrode temperature has a significant effect on breakdown voltage, due to increased electron emission at elevated temperatures. The maximum spark plug temperature, when keeping other parameters constant, is achieved when the lambda-ratio is equal to 0.9, that is, the value for maximum power output. Under lean, and to a lesser extent, rich mixture conditions, the voltage rises; this is largely due to a reduction in the heat released by combustion. Given a constant set of engine operating conditions, an increase in lambda-ratio results in an increased pressure at ignition. This has been attributed to an increase in the ratio of specific heats (the gamma-ratio) of the air-fuel mixture; an increase in gas pressure results in a consequent rise in breakdown voltage [10]-[12].

Changes in lambda-ratio, and therefore in breakdown voltage, lead to subtle changes in the overall shape of the ignition spark waveform. Given a constant ignition system energy, an increase in breakdown voltage results in more energy being used within the breakdown phase. This leaves less energy available for following phases of the spark, i.e., the arc and glow discharge phases. The observed result is a reduction in the glow-discharge duration.

However, factors other than change in lambda are also likely to have an effect on the spark-voltage characteristic. For example, the temperature and pressure inside the cylinder, both of which are related to load, are relevant. In addition, the speed of the engine will determine the degree of in-cylinder turbulence which will also have an effect. Thus, if the voltage characteristic of the spark is to be used to determine the lambda-value, the effects of other parameters also must be accommodated.

To summarize, changes in the value of the lambda-ratio would be expected to influence both the breakdown voltage and the time-varying voltage characteristic of the arc and glow discharge phases. A formal relationship between the value of lambda and the instantaneous voltage at any particular point on the spark-voltage characteristic is not easily discernible and may not exist. However, theoretical considerations indicate a possible correlation between the vector formed by periodic sampling of the voltage at the spark plug over the spark time, termed the *spark-voltage vector*, and the lambda-ratio. With suitable pre-processing and training, a neural network is a suitable tool for associating the spark-voltage vector and lambda-ratio. This forms the basis of the Spark Voltage Characterization technique.

7   Neural Networks for Use in Virtual Sensors

Neural networks possess a number of specific qualities which make them invaluable in pattern-recognition applications and which are not easily achieved by other means. Some of the important qualities of neural networks can be summarized as follows:

- They learn by example and can be conditioned to respond correctly to a stimulus.
- They can automatically perform knowledge abstraction and statistical analyses on data which is presented to them and this information becomes encoded into the internal structure of the network.

- They can generalize so as to respond correctly even in the presence of noise or uncertainty making them suitable for use in poor signal-to-noise environments.

The use of neural networks for application to IC engine sensing [3], [13], [14], diagnostic monitoring [14]-[17] and control [18]-[20] is described in the literature, and new papers appear with increasing frequency. The contribution that neural networks can make in this area may be summarized as follows:

- Neural networks can interpret sensory data which is already present, or available at low cost, so as to extract new information.
- Neural networks can be used for the detection of specific *signatures* from new or existing sensors in OBD systems, in order to detect and identify fault conditions.
- Neural networks, and the related technology, *fuzzy systems*, can be valuable in achieving the non-linear mappings necessary for efficient engine-modeling and the implementation of advanced control strategies.

The SVC method makes use of the pattern-recognition abilities of the neural network for the interpretation of spark voltage vectors. The function of the neural network in this application was to categorize voltage vectors presented to it, differentiating between vectors corresponding to different values of lambda. Certain types of neural network are known to possess useful properties in this area, for example, the multi-layer perceptron (MLP). The MLP is essentially a static network, but it is routinely adapted to process dynamic data by the addition of a tapped delay-line. The delay-line is implemented algorithmically in software, forming the Time-Delay Neural Network [21]. It may be considered that the MLP projects n-element vectors, applied to it as inputs, into n-dimensional input space. Vectors belonging to different classes occupy different regions of this input-space. During the back-propagation learning or training process, a training-file containing exemplar vectors is repeatedly presented to the MLP, and it iteratively places hyper-plane partitions in such positions as to separate the classes attributed to the vectors. During the recall or the operational phase, a vector to be classified is presented to the MLP, which categorizes it by determining where the vector lies in n-dimensional space in relation to the hyper-planes [22].

Feed-forward networks with sigmoidal non-linearities, such as the MLP, are very popular in the literature; however, networks which incorporate radially symmetric processing elements are more appropriate for certain classification applications. The Radial Basis Function (RBF) network is a neural classifier devised in its original form by Moody and Darken [23], but developed and enhanced by others [24]. Usually, the hidden layer consists of elements which perform Euclidean distance calculations, each being followed by a Gaussian activation function. A clustering algorithm is used to calculate the appropriate placings for the cluster centers; for example the k-means algorithm is widely used. In its most elementary form the output layer performs a linear summation of the non-linear outputs of the basis function elements. Alternatively, there can be advantages in the use of the basis neurons as a pre-processing layer for a conventional multi-layer feed-forward neural network, for example an MLP. The non-linear transformation effected by the basis neurons can be considered to move input-vectors into a space of a higher dimension. In some circumstances, the vectors are more easily separated in this higher-dimension space, than in space of their intrinsic dimension. In cases where the topology of the input-space is amenable, the use of RBF networks can lead to improved classification ability; benefits can also accrue in terms of shortened convergence times [24].

The neural network architecture best suited to a particular application depends largely on the topology of the input space (and, of course, on the criteria chosen for the comparison). However, the two network paradigms can be briefly compared as follows:

- The MLP achieves a concise division of the input space, with unbounded or open decision regions, using a comparatively small number of hidden neurons. The RBF network forms bounded or closed decision regions, using a much larger number of pattern (hidden) nodes, to provide a more detailed division of the input space.
- The MLP generally attains a higher classification speed, in its operational or recall mode, than a functionally comparable RBF

network. This is due to the more compact representation, which requires fewer hidden neurons.

8   AFR Estimation using Neural Network Spark Voltage Characterization

Here, a Virtual Lambda Sensor for the estimation of in-cylinder air-fuel ratio is described. The system exhibited a similar level of functionality to the conventional lambda sensor, which determines whether the air-fuel ratio is rich, correct or weak. However, the Virtual Lambda Sensor exhibited the advantage that no dedicated hardware sensor was required.

8.1   The Spark Voltage Characterization Method

The Virtual Lambda Sensor employed the spark voltage characterization method. The correlation between the spark-voltage vector and the lambda-ratio, discussed in Section 6, was exploited by training a neural network to associate specific spark voltage vectors with lambda-values. After training, the neural network was able to determine whether the lambda was correct, rich or weak, when it was presented with a spark voltage vector obtained from the engine operating with that mixture strength. It is recognized that factors other than the lambda would also have an effect on the spark-voltage vector, for example, changes in speed, load, etc. However, initially, the effect of these other parameters was ignored, and experiments were conducted under conditions where only the lambda was varied and other parameters were held constant. Later phases of the work will be concerned with accommodating changes in these other engine parameters.

Two stages in the investigation are presented. Firstly, experimental work using a multi-cylinder engine is described. A number of practical problems are identified, which lead to the second stage of the investigation, where a single-cylinder engine is used.

8.2   Neural Network Training Procedure

Figure 4 shows the experimental arrangement that was used. The engine was equipped with a dynamometer which presented the engine with a "dummy" load that could be varied as desired. The resulting load-torque could be measured and the output power calculated. The throttle setting and air-fuel ratio could be manually adjusted. The air-fuel ratio that resulted from this adjustment was measured by an exhaust gas analyzer. The ignition-system was modified by the addition of a high-voltage test-probe at the spark plug to enable the voltage to be measured and recorded.

A current transformer was fitted to the high-tension line to permit the recording of current data. However, no benefit was obtained from the use of current data and so results are not described.

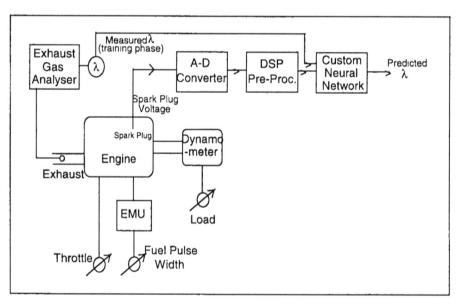

Figure 4. Spark voltage waveform capture system.

An MLP network, with a single hidden layer, and sigmoidal activation units, was used as a spark-voltage vector classifier. The architecture is illustrated in Figure 5. The backpropagation learning algorithm was applied to the MLP during training, which is a supervised training paradigm. This required that the training-file contain spark-voltage vectors, and desired-output vectors. The fuel pulse width and dynamometer were adjusted to give an engine speed and a lambda-ratio of the desired values. Instantaneous spark-voltage vectors of the form $V_n = (v_1, v_2,...,v_n)$ were created by recording the voltage at the spark plug at measured intervals of time. Each spark-voltage vector was associated with a desired-output vector, $D_r = (0,0,1)$, $D_c = (0,1,0)$, and $D_w = (1,0,0)$, depending on whether the lambda-value, measured by the exhaust gas analyzer, was rich, correct or weak, respectively. Three sets of spark-voltage vectors and their associated desired-output vectors were obtained, S_r, S_c and S_w, corresponding to rich, correct and weak lambda values. These vectors were combined into a single training-file, $F = \{S_r \cup S_c \cup S_w\}$. Similar files, having the same construction, but using data that was not used for training, were created for test purposes.

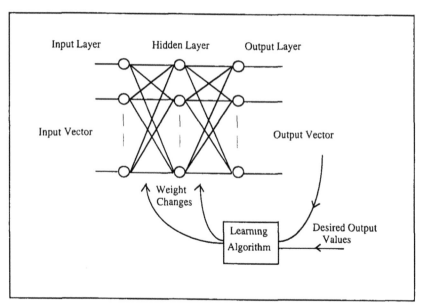

Figure 5. The architecture of the MLP neural network.

The MLP neural network was trained using cumulative back-propagation. The criterion used to determine when the training process should be terminated was based on ensuring that all neuron output values matched the corresponding desired-output value to within a selected convergence threshold T_c. For example, at the termination of the learning phase, the output of the jth output neuron is $y_j \geq 1 - T_c \ \forall \ v_n \in S_j$ and $y_j \leq T_c \ \forall \ v_m \in S_i \ (i = 1,...r) i \neq j$.

8.3   The Multi-Cylinder Engine

An engine test-bed was used that was based on a 1400cc four-cylinder petrol-fueled spark-ignition engine of the type used in many domestic motor-cars. The experiments were conducted at a fixed engine speed of 1500 rpm, with an ignition-timing of 10 degrees before TDC and a wide-open throttle. Stoichiometric, lean and very lean air-fuel ratios were used which corresponded to lambda-ratios of 1.0, 1.2 and 1.4. These values of AFR produced output-torque values of 98.5, 85.0 and 62.8 Nm respectively.

8.3.1   Equal Sample Intervals

Three sets of training-files were constructed. Voltage data were recorded over the full duration of the spark using a fixed sampling interval for each file. The sampling intervals which were used for the three files were 10 µs, 20 µs and 40 µs respectively. Similar files were constructed for testing, using data which was not used for training. An MLP neural network, which executed a custom C-language implementation of the cumulative back-propagation algorithm, was trained using this data. In recall, the test-files were applied to the trained MLP network, where the output of the neural network was modified by a layer which executed a winner-takes-all paradigm. Table 1 shows the performance of the system under these conditions.

Table 1. Correct classification rate for various sampling intervals: single sampling interval.

Sampling Interval (µs)	40	20	10
Correct Classification Rate (%)	71	75	74

8.3.2   Unequal Sample Intervals

A second set of measurements was made with emphasis given to the peak region of the spark by using an increased sampling rate during the peak region compared to that used during tail times. The aim was to capture important transient variations in this region. Three training-files were constructed. Instantaneous voltage measurements were recorded every 2 μs over the peak region for all three files, and then different sampling intervals of 10 μs, 20 μs and 40 μs were used for each file during the remainder of the spark duration. Table 2 shows the results.

Table 2. Correct classification rate for various sampling intervals; peak region emphasized.

Sampling Interval (μs)	40	20	10
Correct Classification Rate (%)	84	80	82

8.3.3   Integration of Instantaneous Values

In an attempt to reduce the effect of the random variations which were observed in successive spark waveforms, integration of instantaneous voltage values over a number of cycles was performed. Different sampling intervals were used during the peak and tail times, as described in Section 8.3.2, and different scale-factors were applied over the two regions. An MLP network was trained using training data which had been pre-processed in this way and a comparison was made using different sizes of training file. The network was trained using files containing 45, 60 and 75 training records, and tested in recall using 45 training records which had not been used in training. Table 3 shows the results that were obtained.

Table 3. Correct classification rate for various numbers of training sets : data integration used and peak region emphasized.

No Training Records	45	60	75
Correct Classification Rate (%)	86	87	93

8.3.4   Radial Basis Functions

In order to investigate whether the use of RBF elements would enhance the classification rate, the data used in Section 8.3.3 were applied to an RBF pre-processing layer, the outputs of which fed an MLP network.

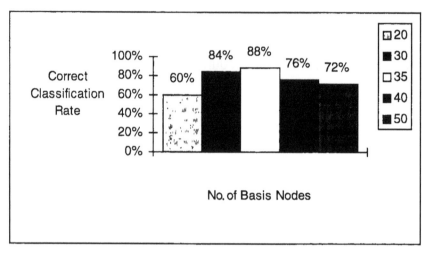

Figure 6. Graph showing correct classification rate for various numbers of hidden/basis nodes: peak region emphasized.

A version of the k-means clustering algorithm, which used semi-supervised learning, was applied to the RBF layer during training. The results which were obtained during recall, when varying numbers of basis nodes were used, are illustrated in Figure 6.

8.3.5 Discussion

Table 1 shows that the neural network could differentiate between the different classes of lambda on the basis of the spark voltage vector with a correct classification rate of between approximately 71 and 75%. With peak region emphasis, an improvement was obtained, as illustrated in Table 2, which shows a correct classification rate of between 80 and 84%. One interpretation of these results is that there was increased information available in the peak region of the spark. As the sampling interval was varied between 10 μs and 40 μs, no significant corresponding variation in classification rate was observed. The results presented in Table 3 show two things: firstly, an improvement in classification rate was observed when integration of instantaneous values was implemented; secondly, further improvements were obtained as the size of the training file was increased. The best classification rate that was obtained under these circumstances was 93%. The best classification rate obtainable using the RBF network was 88%, which was worse than the best rate obtained using the MLP network.

Inspection of the spark waveforms showed that random variations in the shape of successive spark-voltage vectors occurred even when engine parameters were kept as close to constant as practicably possible. The effect could be reduced by the use of integration over successive engine-cycles, as shown by improved results in Table 3. However, this could be an obstacle to the use of this technique for cycle-by-cycle lambda measurement, which is what is ultimately desired.

Observation of the output of the exhaust-gas analyzer showed that there were wide short-term variations in the lambda-ratio, even when the engine parameters were kept as constant as practically possible. These variations could be inherent to engine cyclic variations. A contributory factor could also be that the lambda value that was measured using the exhaust-gas analyzer was an average of the lambda in all four cylinders of the engine. The lambda-value in each cylinder was unlikely to be the same. The recorded spark-voltage vectors were those from only one of these four cylinders. The correlation between the spark-voltage vector obtained from one cylinder and the mean of the lambda-values in all four cylinders was likely to be poor.

The results in Table 3 indicated that better classification was obtained as the size of the training-file increased. However, the inherent instability of the engine made it impossible to maintain constant conditions for the time necessary to collect the required amount of training data.

8.4   The Single-Cylinder Engine

A single-cylinder engine offered a number of advantages over a multi-cylinder power unit. The correlation between the spark voltage vector, measured at the only spark plug of the single cylinder engine, and the lambda measured via the exhaust, was likely to be better than was obtainable in a multi-cylinder unit. The single-cylinder engine would also be likely to offer inherently increased lambda stability, allowing the capture of larger quantities of consistent data, which was required for improved classification.

The experimental arrangement that was used was similar to that shown in Figure 4, the power unit being a single-cylinder four-stroke engine that had a capacity of 98.2cc. The engine was modified to enable manual adjustment to be made to the air-fuel ratio. This was measured using the same exhaust gas composition analyzer as had been used before. The ignition timing was fixed at 24 degrees before top-dead-center. A regenerative electric dynamometer was installed which allowed the load torque to be adjusted to a desired value.

8.4.1   Single-Speed Test

A fixed engine speed of 2800 rpm was selected. Rich, stoichiometric and lean air-fuel ratios were used which corresponded to lambda-ratios of 0.8, 1.0 and 1.2. These values were different to those selected for the multi-cylinder engine, because of the different characteristics of the two power units, but comparable for the purpose of this experiment. The experimental procedure that was described in Section 8.2 was followed. The MLP neural network was trained using a training-file composed of spark-voltage vectors and desired-output vectors. In recall, unseen training data were used. Experiments were conducted with a range of sample intervals. Under these circumstances the neural network virtual-sensor was able to determine the correct lambda-value, 0.8, 1.0 or 1.2 with a correct classification rate of approximately 100%. This performance was superior to that obtained with the multi-cylinder engine, where the best classification rate obtained was 93% (Table 3).

8.4.2   Multi-Speed Tests

A more comprehensive set of tests was carried out on the single-cylinder engine using a more closely spaced range of lambda values, i.e., 0.9, 1.0 and 1.1. A range of speeds and training file sizes was also used. Spark-voltage vectors and desired-output vectors were recorded at speeds of 2800 rpm, 3500 rpm and 4200 rpm. These speed-values corresponded approximately to the lower, middle and upper regions of the working speed range of the engine. Integration over a number of successive cycles was used to reduce the effects of random variations. Three training-files were created, one for each speed. Three similar files, containing data that was not used during training, were constructed for test purposes.

In order to investigate the effects of different numbers of training records, training-files of a number of different sizes were constructed.

The number of records (input-output vector pairs) in the training-file of an MLP network which leads to optimum classification has been the subject of much investigation; however, it has not proved amenable to formal analysis. Investigations described in the literature have indicated that a number of training-records comparable with, or exceeding, the number of weights in the network would lead to good classification ability over a representatively large body of test data. If an MLP network has P, Q and R neurons in the input, hidden and output layers, respectively, the number of weights in the network, N_w, equals $(P + 1)Q + (Q + 1)R$. Letting the number of records in the training file be N_t, then $N_t = \sigma \cdot N_w$ where σ is the *normalized size* of the training file, and $1 < \sigma < 10$ for good classification performance. The optimum value of σ depends on the shape of the P-dimensional feature space, which is, in turn, determined by the problem domain. Generally, large values of σ lead to better classification and generalization; however, adoption of this criterion often leads to a large training-file size and an extended time requirement for network convergence.

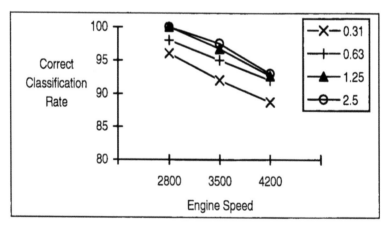

Figure 7. Correct classification rate against engine speed for various normalized training file sizes.

Figure 7 shows the classification performance which was obtained at different speeds and for different values of σ. At 2800 rpm the neural network virtual-sensor could determine the lambda-ratio with a correct classification rate of approximately 100% when either of the largest two file sizes were used during training. Smaller training file sizes resulted in poorer classification rates. At higher speeds the classification rate was not as good. Increasing the size of the training file resulted in an improvement in performance to an extent; however, only a small improvement was evident as the σ is increased from 1.25 to 2.5. This suggested that the decrease in classification ability was due to some inherent change in input data as the speed was increased.

No conclusive reason has been found for the decrease in classification rate with speed and this phenomenon requires further investigation. There are two suggested possible reasons:

- Increased instability in the engine as the speed is increased could result in wider variations in the actual lambda-value about the nominal value. If this was so the signal-to-noise ratio of the data would effectively increase with the speed. This would impair the ability of the neural network to correctly categorize the spark voltage vectors.
- The same sample rate was used for all speeds. At higher speeds fewer measurements were made per revolution. It is possible that the reduced classification rate at higher speeds was due to the worsening of the sampling resolution caused by this.

9   Conclusions

A Virtual Lambda Sensor, using the Spark Voltage Characterization technique, has been introduced here. The system implements neural network analysis of the spark-voltage vector, in order to provide an estimate of the in-cylinder air-fuel ratio. The experimental work shows that the virtual-sensor can provide analogous functionality to the conventional lambda-sensor, but without the need for the usual hardware sensor. The Virtual Lambda Sensor is capable of determining when the lambda-ratio is stoichiometric, or when it deviates from this value by approximately $\pm 10\%$ (*lambda* $= 1.0 \pm 0.1$), with the engine operating under fixed speed and load conditions.

A description has been given of the relatively early stages of the development of the technique. To be practicable as a replacement for the conventional lambda-sensor in a commercial engine, improvements to the Virtual Lambda Sensor are necessary in two respects: firstly, the accuracy of the estimation must be improved, 1% is an aim imposed by the catalytic converter; and secondly, variations in speed, load, etc., must be accommodated.

Improved accuracy demands that the neural network is trained with lambda data which is of higher consistency. Although quantitative measurements are not presented here, observation of the output voltage from the exhaust-gas analyzer using an oscilloscope showed that the lambda-value, under constant engine conditions, could vary from its nominal value by up to approximately 7%. The accuracy which has been achieved is probably close to the best achievable using the current methodology. However, initial results obtained using a more sophisticated experimental methodology have demonstrated improved accuracy.

A mechanism for dealing with variations in speed, load, etc., is the creation of overlays to the neural network weight-matrix for different physical conditions. However, this is likely to impose a large training time penalty. Mathematical analysis of the dynamic physical system is also being implemented to provide guidance about the optimum pre-processing of the data before it is used in the neural network training phase.

Acknowledgment

This work was carried out at the *Transfrontier Centre for Automotive Research* (**TCAR**) which is financially supported by the European Union under the Interreg II Programme of the European Regional Development Fund, grant number ES/B3/01.

References

[1] Kimberley, W. (1997), "Focus on emissions," *Automotive Engineer,* Vol. 22, No.7, pp. 50-64.

[2] Boam, D.J., Finlay, I.C., Biddulph, T.W., Ma, T.A., Lee, R., Richardson, S.H., Bloomfield, J., Green, J.A., Wallace, S., Woods, W.A. and Brown, P. (1994), "The sources of unburnt hydrocarbon emissions from spark ignition engines during cold starts and warm-up," *Proceedings of The Institution of Mechanical Engineers. Journal of Automobile Engineering, Part D*, 208, pp. 1-11.

[3] Atkinson, C.M., Long, T.W. and Hanzevack, E.L. (1998), "Virtual sensing: a neural network-based intelligent performance and emissions prediction system for on-board diagnostics and engine control," *Proceedings of the 1998 SAE International Congress & Exposition*, vol. 1357, pp. 39-51.

[4] Cardini, P. (1999), "Focus on emissions: going for the burn," *Automotive Engineer*, vol. 24, no. 8, pp. 48-52.

[5] Eriksson, L. and Nielsen, L. (1997), "Ionization current interpretation for ignition control in internal combustion engines," *Control Engineering Practice*, vol. 5, no 8, pp. 1107-1113.

[6] Balles, E.N., VanDyne, E.A., Wahl, A.M., Ratton, K. and Lai, M.C. (1998), "In-cylinder air/fuel ratio approximation using spark gap ionization sensing," *Proceedings of the 1998 SAE International Congress & Exposition*, vol. 1356, pp. 39-44.

[7] Ohashi, Y., Fukui, W., Tanabe, F. and Ueda, A. (1998), "The application of ionic current detection system for the combustion limit control," *Proceedings of the 1998 SAE International Congress & Exposition*, vol. 1356, pp. 79-85.

[8] Hellring, M., Munther, T., Rognvaldsson, T., Wickstrom, N., Carlsson, C., Larsson, M., and Nytomt, J. (1998), "Spark Advance Control using the Ion Current and Neural Soft Sensors," *SAE Paper 99P-78*.

[9] Hellring, M., Munther, T., Rognvaldsson, T., Wickstrom, N., Carlsson, C., Larsson, M. and Nytomt, J. (1998), "Robust AFR estimation using the ion current and neural networks," *SAE Paper 99P-76.*

[10] Champion Spark Plugs. (1987), *Straight Talk About Spark Plugs.*

[11] NGK Spark Plug Co. Ltd. (1991), *Engineering Manual For Spark Plugs*, OP-0076-9105.

[12] Pashley, N.C. (1997), *Ignition Systems For Lean-burn Gas Engines*, Ph.D. Thesis, Department of Engineering Science, University of Oxford, U.K.

[13] Frith, A.M., Gent, C.R. and Beaumont, A.J. (1995), "Adaptive control of gasoline engine air-fuel ratio using artificial neural networks," *Proceedings of the Fourth International Conference on Artificial Neural Networks*, no. 409, pp. 274-278.

[14] Ayeb, M., Lichtenthaler, D., Winsel, T. and Theuerkauf, H.J. (1998), "SI engine modeling using neural networks," *Proceedings of the 1998 SAE International Congress & Exposition*, vol. 1357, pp. 107-115.

[15] Wu, Z.J. and Lee, A. (1998), "Misfire detection using a dynamic neural network with output feedback," *Proceedings of the 1998 SAE International Congress & Exposition*, vol. 1357, pp. 33-37.

[16] Ribbens, W.B., Park, J., and Kim, D. (1994), "Application of neural networks to detecting misfire in automotive engines," *IEEE International Conference on Acoustics, Speech and Signal Processing*, vol. 2, pp. 593-596.

[17] Ortmann, S., Rychetsky, M., Glesner, M., Groppo, R., Tubetti, P. and Morra, G. (1998), "Engine knock estimation using neural networks based on a real-world database," *Proceedings of the 1998 SAE International Congress & Exposition*, vol. 1357, pp. 17-24.

[18] Muller, R. and Hemberger, H.H. (1998), "Neural adaptive ignition control," *Proceedings of the 1998 SAE International Congress & Exposition*, vol. 1356, pp. 97-102.

[19] Lenz, U. and Schroder, D. (1998), "Air-fuel ratio control for direct injecting combustion engines using neural networks," *Proceedings of the 1998 SAE International Congress & Exposition*, vol. 1356, pp. 117-123.

[20] Baumann, B., Rizzoni, G. and Washington, G. (1998), "Intelligent control of hybrid vehicles using neural networks and fuzzy logic," *Proceedings of the 1998 SAE International Congress & Exposition*, vol. 1356, pp. 125-133.

[21] Hush, D.R. and Horne, B.G. (1993), "Progress in supervised neural networks," *IEEE Signal Processing Magazine*, pp. 8-39.

[22] Hush, D.R., Horne, B. and Salas, J.M. (1992), "Error surfaces for multilayer perceptrons," *IEEE Transactions on Systems, Man and Cybernetics*, vol.22, no.5, pp.1152-1161.

[23] Moody, J. and Darken, C.J. (1989), "Fast learning in networks of locally tuned processing units," *Neural Computation*, vol.1, pp. 281-294.

[24] Leonard, J.A., Kramer, M.A. and Ungar, L.H. (1992), "Using radial basis functions to approximate a function and its error bounds" *IEEE Transactions on Neural Networks*, vol.3, no.4, pp. 625-627.

[25] Fu, L. (1994), *Neural Networks in Computer Intelligence*, McGraw-Hill, New York.

CHAPTER 10

NEURAL ARCHITECTURES OF FUZZY PETRI NETS

W. Pedrycz
Department of Electrical & Computer Engineering
University of Alberta, Edmonton
Canada T6G 2G7
&
Systems Research Institute
Polish Academy of Sciences
Warsaw, Poland
pedrycz@ee.ualberta.ca

In this chapter, we discuss a novel approach to pattern classification using a concept of fuzzy Petri nets. In contrast to the commonly encountered Petri nets with their inherently Boolean character of processing tokens and firing transitions, the proposed generalization involves continuous variables. This extension makes the nets to be fully in rapport with the panoply of the real-world classification problems. The introduced model of the fuzzy Petri net hinges on the logic nature of the operations governing its underlying behavior. The logic-driven effect in these nets becomes especially apparent when we are concerned with the modeling of its transitions and expressing pertinent mechanisms of a continuous rather than an on-off firing phenomenon. An interpretation of fuzzy Petri nets in the setting of pattern classification is provided. This interpretation helps us gain a better insight into the mechanisms of the overall classification process. Input places correspond to the features of the patterns. Transitions build aggregates of the generic features giving rise to their logical summarization. The output places map themselves onto the classes of the patterns while the marking of the places correspond to the class of membership values. Details of the learning algorithm are also provided along with an illustrative numeric experiment.

0-8493-2268-5/2000/$0.00+$ 50
© 2000 by CRC Press LLC

1   Introduction

In recent years, Petri nets [5] have started to gain in importance in the areas of knowledge representation, robot planning, and expert systems, see, for instance, [1], [2], [3], and [4]. Surprisingly, little research has been done on the use of Petri net in pattern classification. On the other hand, most classification pursuits are easily formalized in the setting of Petri nets once these architectures become generalized in a way that they reflect a continuous character omnipresent in most of the classification tasks. The approach taken here dwells on the existing fuzzy set-based augmentation of the generic version of Petri nets [6], [7]. Fuzzy sets contribute not only to a way in which an issue of partial firing of the transitions can be addressed but they provide a significant level of parametric flexibility. This flexibility becomes indispensable in the case of training of such fuzzy Petri nets – the feature not being available in their standard binary counterparts. The objective of this study is to investigate fuzzy Petri nets in the framework of pattern recognition and make them conceptually and computationally viable as pattern classifiers.

The material of the chapter is arranged into 7 sections. We start off with a brief introduction to Petri nets along with their fuzzy set-based generalization. This generalization helps us capture and formalize the notion of continuous rather than straight on-off firing mechanism. In Section 3, we analyze the details of the fuzzy Petri net providing all necessary computational details. Subsequently, Section 4 deals with a process of learning in the nets. The main objective of such learning is to carry out some parametric optimization so that the network can adjust to the required training set of patterns (being composed of pairs of marking of input and output places). A way of interfacing fuzzy Petri nets with the modeling environment is discussed in Section 5. Numerical experiments are reported in Section 6 while conclusions are covered in Section 7.

2 The Generalization of the Petri Net and Its Underlying Architecture

Let us briefly recall the basic concept of a Petri net. Formally speaking, a Petri net [4], [5] is a finite graph with two types of nodes, known as places (P) and transitions (T). More formally, the net can be viewed as a triple (P, T, F) where

$$P \cap T = \varnothing$$

$$P \cup T \neq \varnothing$$

$$F \subseteq (P \times T) \cup (T \times P)$$

$$\text{domain}(F) \cup \text{codomain}(F) = P \cup T$$

In the above, F is called the flow relation. The elements of F are the arcs of the Petri net.

Each place comes equipped with some tokens that form a marking of the Petri net. The flow of tokens in the net occurs through firings of the transitions; once all input places of a given transition have a nonzero number of tokens, this transition fires. Subsequently, the tokens are allocated to the output places of the transition. Simultaneously, the number of tokens at the input places is reduced. The effect of firing of the transitions is binary: the transition either fires or does not fire.

An important generalization of the generic model of the Petri net is to relax the Boolean character of the firing process of the transition. Instead of subscribing to the firing-no firing dichotomy, we propose to view the firing mechanism as a gradual process with a continuum of possible numeric values of the strength (intensity) of firing of a given transition. Then the flow of tokens can also take this continuity into consideration meaning that we end up with the marking of the places that become continuous as well. Evidently, such a model is in rapport with a broad class of real-world phenomena including pattern classification. The generalization of the net along this line calls for a series of pertinent realization details. In what follows, we propose a construct whose functioning adheres as much as possible to the logic fabric delivered by fuzzy sets. In this case, a sound solution is to adopt the ideas of fuzzy logic as the most direct way of implementation of such networks.

3   The Architecture of the Fuzzy Petri Net

The topology of the fuzzy Petri net as being cast in the framework of pattern classification is portrayed in Figure 1. As it will be shown further on, this setting nicely correlates with the classification activities encountered in any process of pattern recognition.

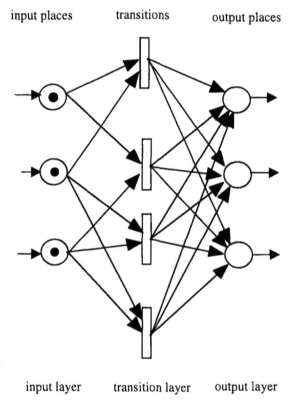

Figure 1. A general three layer topology of the fuzzy Petri net.

The network constructed in this manner comprises three layers:

- an input layer composed of "n" input places;

- a transition layer composed of "hidden" transitions;

- an output layer consisting of "m" output places.

The input place is marked by the value of the feature (we assume that the range of the values of each of the features is in the unit interval). These marking levels are processed by the transitions of the network whose levels of firing depend on the parameters associated with each transition such as their threshold values and the weights (connections) of the incoming features. Subsequently, each output place corresponds to a class of patterns distinguished in the problem. The marking of this output place reflects a level of membership of the pattern in the corresponding class.

The detailed formulas of the transitions and output places rely on the logic operations encountered in the theory of fuzzy sets. The i-th transition (more precisely, its activation level z_i) is governed by the expression

$$z_i = \underset{j=1}{\overset{n}{T}}[w_{ij}s(r_{ij} \rightarrow x_j)], \quad j = 1, 2, ..., n; i = 1, 2, ..., \text{hidden},$$

where:

- w_{ij} is a weight (connection) between the i-th transition and the j-th input place;

- r_{ij} is a threshold level associated with the level of marking of the j-th input place and the i-th transition; and

- the level of marking of the j-th input place is denoted by x_j

Moreover, "t" is a t-norm, "s" denotes an s-norm, while $\rightarrow$ stands for an implication operation expressed in the form

$$a \rightarrow b = \sup\{c \in [0,1] \mid atc \leq b\} \qquad (1)$$

where a, b are the arguments of the implication operator confined to the unit interval. Note that the implication is induced by a certain t-norm. In the case of two-valued logic, (1) returns the same truth values as the commonly known implication operator, namely

$$a \rightarrow b = \begin{cases} b, \text{if } a > b \\ 1, \text{otherwise} \end{cases} = \begin{cases} 0, \text{if } a = 1 \text{ and } b = 0 \\ 1, \text{otherwise} \end{cases} \quad a, b \in \{0,1\}$$

The j-th output place (more precisely, its marking y_j) summarizes the levels of evidence produced by the transition layer and performs a nonlinear mapping of the weighted sum of the activation levels of these transitions (z_i) and the associated connections v_{ji}

$$y_j = f(\sum_{i=1}^{hidden} v_{ji} z_i), \qquad j = 1, 2, ..., m \tag{2}$$

where "f" is a nonlinear monotonically increasing function (mapping) from $\mathbf{R}$ to $[0,1]$.

The role of the connections of the output places is to modulate an impact the firing of the individual transition exhibits on the accumulation of the tokens at this output place (viz. the membership value of the respective class). The negative values of the connections have an inhibitory effect meaning that the value of the class membership becomes reduced.

Owing to the type of the aggregation operations being used in the realization of the transitions, their interpretation sheds light on the way in which the individual features of the problem are treated. In essence, the transition produces some higher level, synthetic features out of these originally encountered in the problem and represented in the form of the input places. The weight (connection) expresses a global contribution of the j-th feature to the i-th transition: the lower the value of w_{ij}, the more significant the contribution of the feature to the formation of the synthetic aggregate feature formed at the transition level. The connection itself is weighted uniformly regardless of the numeric values it assumes. The more selective (refined) aggregation mechanism is used when considering threshold values. Referring to (1), one easily finds that the thresholding operation returns 1 if x_j exceeds the value of the threshold r_{ij}. In other words, depending on this level of the threshold, the level of marking of the input place becomes "masked" and the threshold operation returns one. For the lower values of the marking, such levels are processed by the implication operation and contribute to the overall level of the firing of the transition.

One should emphasize that the generalization of the Petri net proposed here is in full agreement with the two-valued generic version of the net commonly encountered in the literature. Consider, for instance, a single transition (transition node). Let all its connections and thresholds be restricted to $\{0, 1\}$. Similarly, the marking of the input places is also quantified in a binary way. Then the following observations are valid:

- only those input places are relevant to the functioning of the i-th transition for which the corresponding connections are set to zero and whose thresholds are equal to 1. Denote a family of these places by **P**

- the firing level of the i-th transition is described by the following formula,

$$z_i = \mathop{T}_{j \in P}^{n} (r_{ij} \rightarrow x_j)$$

It becomes apparent that the firing level is equal to 1 if and only if the marking of all input places in **P** assume 1; the above expression for the transition is nothing but an *and*-combination of the levels of marking of the places in **P**, namely

$$z_i = \mathop{T}_{j \in P}^{n} x_j$$

(let us recall that any t-norm can be used to model the *and* operation; moreover all t-norms are equivalent when operating on the 0-1 truth values).

4   The Learning Procedure

The learning completed in the fuzzy Petri net is the one of parametric nature, meaning that it focuses on changes (updates) of the parameters (connections and thresholds) of the net. The structure of the net is left unchanged. These updates are carried in such a way so that a certain predefined performance index becomes minimized. To help concentrate on the detailed derivation of the learning formulas, it is advantageous to view a fully annotated portion of the network as illustrated in Figure 2.

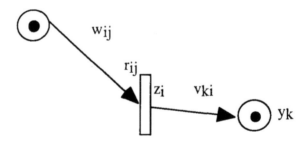

input layer   transition layer   output layer

Figure 2. Optimization in the fuzzy Petri net; a section of the net outlines all notation being used in the learning algorithm.

The performance index to be minimized is viewed as a standard sum of squared errors. The errors are expressed as differences between the levels of marking of the output places of the network and their target values. The considered on-line learning assumes that the modifications to the parameters of the transitions and output places occur after presenting an individual pair of the training sets, say marking of the input places (denoted by $\mathbf{x}$) and the target values (namely, the required marking of the output places) expressed by $\mathbf{t}$. Then the performance index for the input-output pair reads as

$$Q = \sum_{k=1}^{m} (t_k - y_k)^2 \tag{3}$$

The updates of the connections are governed by the standard gradient-based method

$$\mathbf{param}(\text{iter} + 1) = \mathbf{param}(\text{iter}) - \alpha \nabla_{\text{param}} Q \tag{4}$$

where $\nabla_{\text{param}} Q$ is a gradient of the performance index Q taken with respect to the parameters of the fuzzy Petri net. The iterative character of the learning scheme is underlined by the parameter vector regarded as a function of successive learning epochs (iterations).

The intensity of learning is controlled by the positive learning rate (α). In the above scheme, the vector of the parameters, **param**, is used to encapsulate all the elements of the structure to be optimized. Further on they will be made more detailed, as we will proceed with a complete description of the update mechanism. With the quadratic performance index (3) in mind, the following holds

$$\nabla_{\text{param}} Q = -2 \sum_{k=1}^{m} (t_k - y_k) \, \nabla_{\text{param}} y_k$$

Moving into detailed formulas refer again to Figure 2. Moreover, the nonlinear function associated with the output place (2) is a standard sigmoid nonlinearity described as

$$y_k = \frac{1}{1 + \exp(-z_k)}$$

For the connections of the output places we obtain

$$\frac{\partial y_k}{\partial v_{ki}} = y_k (1 - y_k) z_i$$

k=1, 2, ...,m, i=1, 2, ..., hidden. Observe that the derivative of the sigmoidal function is equal to $y_k(1 - y_k)$.

Similarly, the updates of the threshold levels of the transitions of the net are expressed in the form

$$\frac{\partial y_k}{\partial r_{ij}} = \frac{\partial y_k}{\partial z_i} \frac{\partial z_i}{\partial r_{ij}} \qquad i=1, 2, ...,n.$$

In the sequel, we obtain

$$\frac{\partial y_k}{\partial z_i} = y_k (1 - y_k) v_{ki}$$

and

$$\frac{\partial z_i}{\partial r_{ij}} = A \frac{\partial}{\partial r_{ij}} (w_{ij} + (r_{ij} \rightarrow x_j) - w_{ij}(r_{ij} \rightarrow x_j)) = A(1 - w_{ij}) \frac{\partial}{\partial r_{ij}} (r_{ij} \rightarrow x_j)$$

where the new expression, denoted by A, is defined by taking the t-norm over all the arguments but "j",

$$A = \mathop{T}_{\substack{l=1 \\ l \neq j}}^{n} [w_{il} s(r_{il} \rightarrow x_l)]$$

The calculations of the derivative of the implication operation can be completed once we confine ourselves to some specific realization of the t-norm that is involved in its development. For the product (being a particular example of the t-norm), the detailed result reads as

$$\frac{\partial}{\partial r_{ij}}(r_{ij} \rightarrow x_j) = \frac{\partial}{\partial r_{ij}} \begin{cases} x_j / r_{ij} &, \text{if } r_{ij} > x_j \\ 1, \text{otherwise} \end{cases} = \begin{cases} -x_j / r^2_{ij} &, \text{if } r_{ij} > x_j \\ 0, \text{otherwise} \end{cases}$$

The derivatives of the connections of the transitions (transition nodes) are obtained in a similar way. We get

$$\frac{\partial y_k}{\partial w_{ij}} = \frac{\partial y_k}{\partial z_i} \frac{\partial z_i}{\partial w_{ij}} \qquad k=1, 2, \ldots, m, \ i=1, 2, \ldots, \text{hidden}, \ j=1, 2, \ldots, n$$

Subsequently, one derives

$$\frac{\partial z_i}{\partial w_{ij}} = A\frac{\partial}{\partial w_{ij}}(w_{ij} + (r_{ij} \rightarrow x_j) - w_{ij}(r_{ij} \rightarrow x_j)) = A(1-(r_{ij} \rightarrow x_j))$$

There are two aspects of further optimization of the fuzzy Petri nets that need to be raised in the context of their learning:

- the number of nodes in the transition layer. The optimization of the number of the transition nodes of the fuzzy Petri net falls under the category of structural optimization that cannot be handled by the gradient-based mechanisms of the parametric learning. By increasing the number of these nodes, we enhance the mapping properties of the net as each transition can be fine-tuned to fully reflect the boundaries between the classes. Too many of these transitions, however, could easily develop a memorization effect that is well-known in neural networks.

- the choice of specific t-norm and s-norm. This leads us to an aspect of a semi-parametric optimization of the fuzzy Petri net. The choice of these norms does not impact the architecture of the net; yet in this optimization we cannot resort ourselves to the gradient-based learning. A prudent way to follow would be to confine to a family of t-norms and s-norms that can be systematically exploited one by one.

5   Interfacing Fuzzy Petri Nets with Granular Information

In addition to the design of the fuzzy Petri net, we are also concerned with its interfacing of the environment in which it has to perform. It is accomplished by defining a certain functional module that helps transform physical entities coming from the environment into more abstract and logic-inclined entities to be used directly by the fuzzy Petri net. The essence of such interfaces is in information granulation and the use of such granules as a bridge between a numeric information generated by the physical environment and the logical layer of information granules visible at the level of the Petri net itself. In what follows, we elaborate on these two phases in more detail:

(i) *Construction of information granules.* The granulation of information can be carried out in different ways. The use of Fuzzy C-Means or other fuzzy clustering technique is a viable way to follow. Let us briefly recall that the crux of clustering is to form information granules – fuzzy sets (fuzzy relations) – when starting from a cloud of numeric data. The result of clustering comes in the form of the prototypes of the clusters and a fuzzy partition summarizing a way the entire data set becomes divided (split) into clusters. As the name stipulates, the partitioning assigns elements to different clusters to some degree with the membership values between 0 and 1. For "c" clusters we end up with "c" prototypes, $v_1, v_2, \ldots, v_c$.

(ii) *Determination of the activation levels of information granules.* Any new input x "activates" the i-th cluster according to the formula

$$u_i(\mathbf{x}) = \cfrac{1}{\displaystyle\sum_{i=1}^{c} \cfrac{\|\mathbf{x} - \mathbf{v}_i\|^2}{\|\mathbf{x} - \mathbf{v}_j\|^2}}$$

where ∥.∥ is a distance function defined between $\mathbf{x}$ and the respective prototype. The calculations shown above form a core of the interface structure of the fuzzy Petri net, see Figure 3. The activation levels are directly used as the marking of the input places.

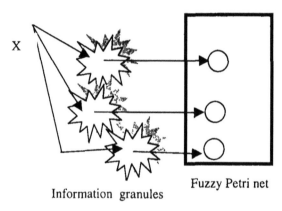

Figure 3. Interfacing the world with the fuzzy Petri net.

An important issue arises when the input ($\mathbf{x}$) is not numeric but comes as some less specific information granule. In particular, we may anticipate a granular information represented in the form of a certain numeric interval, see Figure 4.

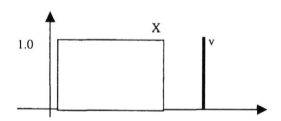

Figure 4. An interval-based information granule and distance calculations.

Taking into consideration the interval-based information granule, the distance function needs some modification. The calculations are carried out in the form

$$\|X - v_i\| = \min_{z \in X} \|z - v_i\|$$

where X is the numeric granule. (Note that the above formula deals with a certain coordinate of **x** and prototype **v**, say x and v.) These computations provide us with an optimistic (that is minimal) value of the distance function. This, in turn, activates the linguistic granules to a higher extent in comparison to what it would have been in the pure numeric case. Subsequently, the activation levels sum up to the value higher than 1 (recall that this sum of the activation levels is always equal to one in the case of numeric inputs coming from the modeling environment). In this sense this sum (or its departure from one) can serve as a useful indicator of the granularity level of the input information.

To quantify this observation, we discuss a one-dimensional case and consider three prototypes located at 1.0, 4.0, and 7.0, respectively. The nonnumeric information is represented as an interval distributed around x with the bounds located at x–d and x+d. The distance is computed as discussed above. The plots of the activation of the first information granule distributed around 1.0 are illustrated in Figure 5.

Subsequently, Figure 6 visualizes a sum of the activation levels of the linguistic granules implied by the interval-valued input. Observe that with the increase of "d", the sum starts to exceed 1.

6   Experiments

In this section, we discuss some experimental results. For illustrative purposes we focus on the classification problems that involve only two features.

Experiment 1.   The patterns themselves are generated by two fuzzy functions

$$f_1(x_1, x_2) = (\overline{x_1}tx_2)s(x_1t\overline{x}_2)$$

$$f_2(x_1, x_2) = (x_1t0.2)s(\overline{x}_2) \tag{5}$$

In these two functions, the t- and s-norms are implemented using the product (atb=ab) and probabilistic sum (asb=a+b–ab), respectively, $a, b \in [0,1]$. The overbar denotes a complement of the feature's value,

$\bar{x} = 1 - x$, $x \in [0,1]$. Note that the first function is a multivalued (fuzzy) Exclusive-Or function (XOR) whereas the second one assumes its high truth values at the upper corner of the unit square. The values assumed by these two functions are treated as continuous class memberships. The two variables of the functions form the features. The plots contained in Figures 7 and 8 portray the functions themselves as well as the classification boundaries occurring between the classes (viz. the curves along which the membership grades in two classes are equal).

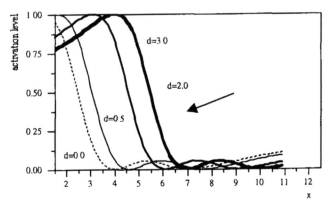

Figure 5. Activation level (membership grade) of the first prototype regarded as a function of x for selected values of information granularity (d).

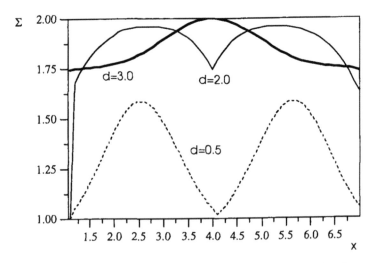

Figure 6. Sum of activation levels (Σ) of the linguistic granules for some selected values of "d".

Neural Architectures of Fuzzy Petri Nets

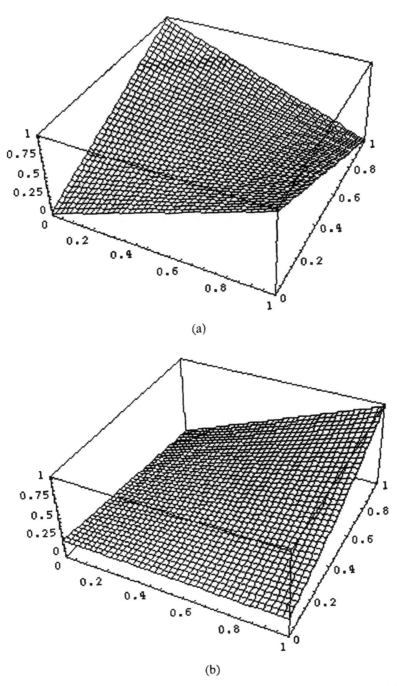

(a)

(b)

Figure 7. Plots of the two logic functions used in the experiment (a) f_1 and (b) f_2

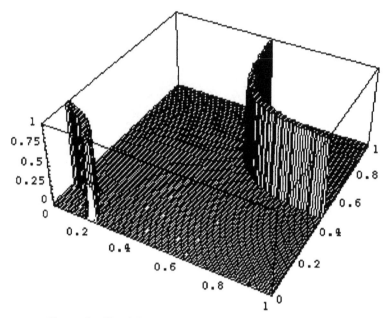

Figure 8. Classification boundaries between the classes.

Note that the problem itself is nonlinear; furthermore the classification boundaries give rise to the disjoint classification regions.

Let us now set up a topology of the fuzzy Petri net. The network has four input places. These correspond to the two original features and their complements. The output layer consists of two output places reflecting the number of the classes occurring in the classification problem. The number of transitions (viz. the size of the transition layer, denoted by "hidden") varies from 2 to 8 throughout the topologies of the Petri nets. The learning was completed in an on-line version meaning that the updates of the connections (parameters) of the Petri net are carried out for each input-output pair of the training data. The experiment involves a standard training-testing scenario: the training was done based on 100 patterns generated randomly (over $[0,1] \times [0,1]$) from the model (5); the testing set involves another 100 patterns again governed by (5). The corresponding data sets are illustrated in Figures 9 and 10.

The learning rate (α) was set to 0.02; the intent was to assure a stable learning process not allowing for any oscillations. Obviously, one can increase the value of the learning rate and therefore accelerate learning without sacrificing its stability. Nevertheless, the issue of efficiency of learning was not a primary concern in this experiment. This is particularly so, as the learning itself is rather fast and does not require a significant number of learning epochs.

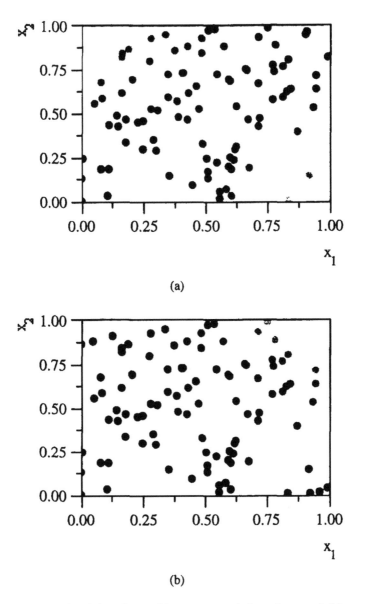

(a)

(b)

Figure 9. A set of training data (100 patterns): (a) first class, and (b) second class; the darker the pattern, the lower its class membership grade.

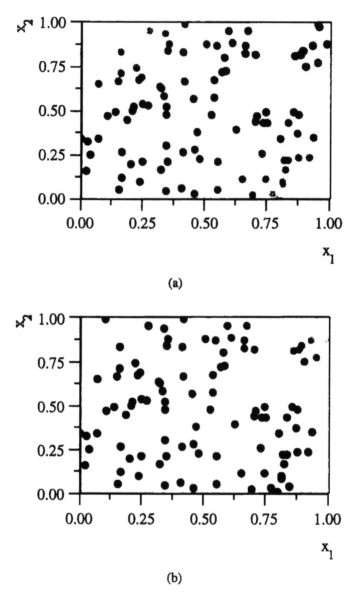

Figure 10. A set of testing data (100 patterns): (a) first class, and (b) second class; the darker the pattern, the lower its class membership grade.

The learning results summarized in the form of the performance index are provided in Table 1. They show the behavior of the fuzzy Petri net on the training set vis-à-vis the results obtained for the testing set. Several conclusions can be drawn from these results:

Table 1. Performance of the fuzzy Petri net (training and testing set) for various number of the transition nodes.

number of transition nodes	2	3	4	5	6	7	8
training set	2.367	0.140	0.079	0.069	0.070	0.050	0.062
testing set	2.155	0.121	0.069	0.067	0.053	0.043	0.073

- It becomes apparent that the number of transition nodes equal to 4 gives rise to a useful architecture that is not excessively large and still produces good classification results. Going toward a higher number of the transition nodes and eventually accepting an excessive size of the network does not yield a significant decrease in the values of the optimized performance index.

- There is an apparent jump in the performance of the network equipped with two transitions and the other versions of the net equipped with three or more transitions.

The following series of figures, Figures 11 to 13, illustrate more details dealing with the learning and performance of the fuzzy Petri net. This concerns a way in which the learning proceeds, visualizes the resulting firing levels of the transitions of the net, and illustrates the values of the classification errors reported for the individual patterns.

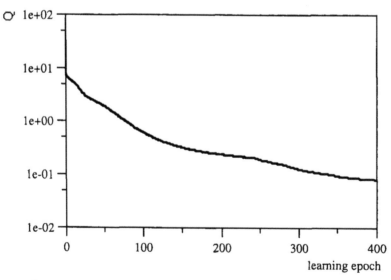

Figure 11. Performance index Q in successive learning epochs.

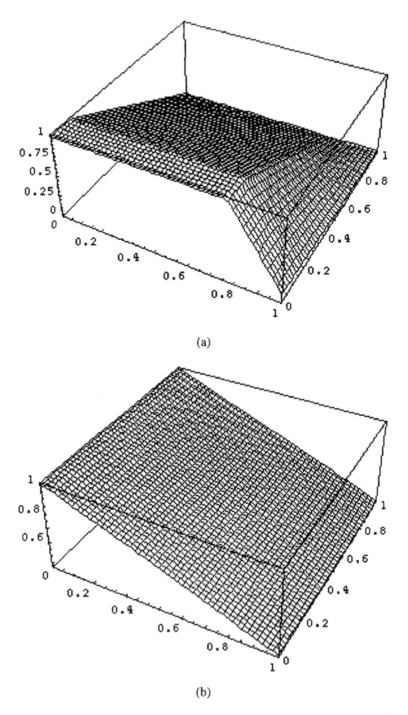

(a)

(b)

Figure 12.   Characteristics of the transitions (transition nodes) regarded as
functions of input variables x_1 and x_2 (continued on next page).

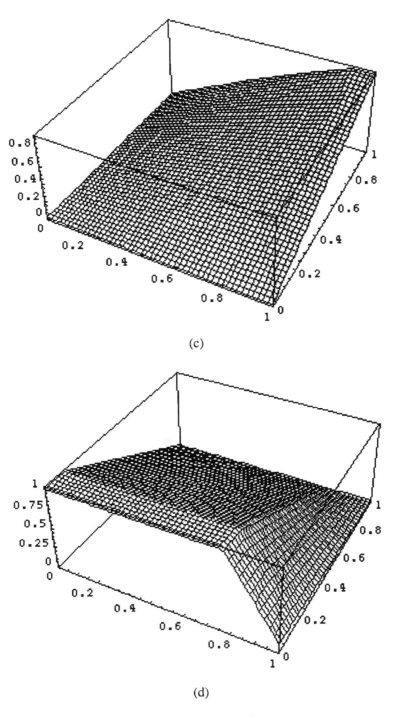

(c)

(d)

Figure 12 (continued).

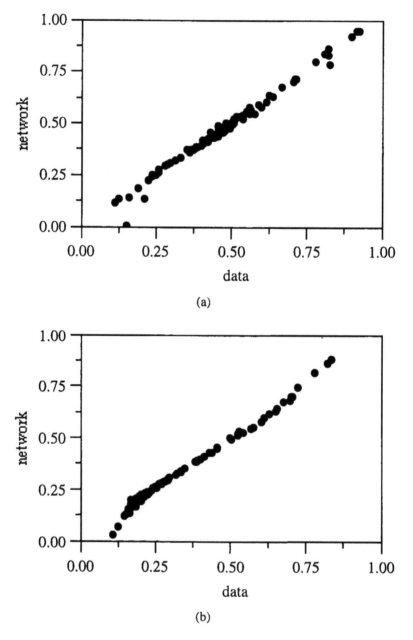

Figure 13. Results of the network and the data: (a) first output; (b) second output.

The connections of the classifier are provided in the form of the following matrices (that are the connections and thresholds of the fuzzy Petri net):

$$W = \begin{bmatrix} 0.9843 & 0.9968 & 0.0838 & 0.0066 \\ 0.9956 & 0.7385 & 0.4364 & 0.9986 \\ 0.0016 & 0.0332 & 0.9031 & 0.9950 \\ 0.0018 & 0.3866 & 0.6169 & 0.2966 \end{bmatrix}$$

$$R = \begin{bmatrix} 0.0000 & 0.2674 & 0.2364 & 0.8721 \\ 0.5869 & 0.0000 & 0.9722 & 0.7987 \\ 0.9106 & 0.9763 & 0.6158 & 0.9749 \\ 0.7429 & 0.0000 & 0.2157 & 0.8697 \end{bmatrix}$$

$$V = \begin{bmatrix} -3.8568 & 2.1462 & -4.3472 & 3.6591 \\ 0.6068 & -2.1980 & 3.2047 & -0.8616 \end{bmatrix}$$

<u>Experiment 2</u>. Here we study two other two-variable fuzzy functions governed by the expressions

$$f_1(x_1, x_2) = [(0.3sx_1)(1 - x_2)]s[(1 - x_1)x_2]$$

$$f_2(x_1, x_2) = [x_1 x_2]s[(1 - x_2)(0.4s(1 - x_1))]$$

As a matter of fact, these give rise to the generalization of the exclusive-OR problem. The functions are also shown in Figure 14. The class boundaries clearly underline the nonlinear character of the classification problem, see Figure 15.

The results of learning for different sizes of the hidden layer (that is the number of transitions) are summarized in Table 2. Apparently the minimized error becomes significantly reduced at h = 5 and afterwards remains fairly stable (this effect is visible for both the learning and testing set).

Table 2. Performance of the fuzzy Petri net (both training and testing set) for various number of the transition nodes.

number of transition nodes	2	3	4	5	6	7	8
learning set	4.6357	0.5376	0.3087	0.0415	0.0424	0.0487	0.0463
testing set	6.0824	0.6922	0.4991	0.0986	0.1169	0.1057	0.0976

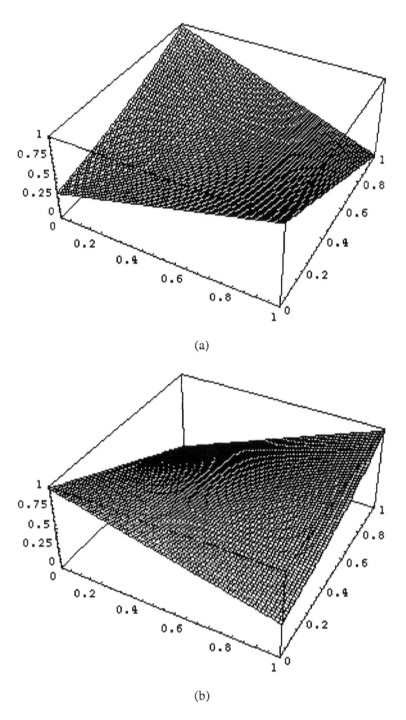

(a)

(b)

Figure 14. 3-D plots of the two-variable logic functions, f_1(a) and f_2(b).

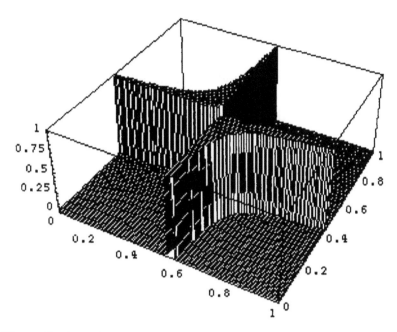

Figure 15. Classification boundaries in the two-class classification problem.

7   Conclusions

In this chapter, we have proposed a new approach to pattern classification, dwelling on the concept of the fuzzy Petri net. Two features of this architecture are definitely worth underlining. The first one concerns a transparent form of the classification model where each component of the fuzzy Petri net (places and transitions) comes as a clearly defined functional entity. The elements in the transition layer give rise to the combination of the original features thus producing new aggregates (synthetic features). The output places are used to aggregate evidence about class membership. Secondly, the Petri network exhibits a high level of parametric flexibility by coming equipped with a significant number of adjustable parameters (such as threshold levels of the transitions and the connections of the transition nodes as well as the output places).

The complete learning scheme has been proposed and illustrated with the aid of numeric examples. While the experiments dealt primarily with some specific forms of the t- and s-norms, it would be advisable to experiment with a wide range of such logic operators and view this as an extra component of flexibility available in the design of such generalized Petri nets. The neuro-like style of performance of the proposed Petri net model being applied to classification problems provides us with a different and definitely interesting insight into the classification activities that is primarily based on features viewed as important resources utilized toward pattern classification.

The study has laid down the fundamentals of the new and general pattern recognition scheme. More specific application areas worth revisiting in this setting deal with scene analysis and computer vision where one can easily encounter parallel threads of classification pursuits.

Acknowledgment

Support from the Natural Sciences and Engineering Research Council of Canada (NSERC) is gratefully acknowledged.

References

[1] Cao, W.T. and Sanderson, A.C. (1995), "Task sequence planning using fuzzy Petri nets," *IEEE Trans. on Systems, Man, and Cybernetics*, vol. 25, pp. 755-768.

[2] Garg, M.L., Ahson, S.I., and Gupta, P.V. (1991), "A fuzzy Petri net for knowledge representation and reasoning," *Information Processing Letters*, vol. 39, pp. 165-171.

[3] Konar, A. and Mandal, A.K. (1996), "Uncertainty management in expert systems using fuzzy Petri nets," *IEEE Trans. on Knowledge and Data Engineering*, vol. 8, pp. 96-105.

[4] Looney, C.G. (1988), "Fuzzy Petri nets for rule-based decision making," *IEEE Trans. on Systems, Man, and Cybernetics*, vol. 18, pp. 178-183.

[5] Murata, T. (1989), "Petri nets: properties, analysis, and applications," *Proc. of the IEEE*, vol. 77, pp. 541-580.

[6] Pedrycz, W. and Gomide, F. (1994), "A generalized fuzzy Petri net model," *IEEE Trans. on Fuzzy Systems*, vol. 2, pp. 295-301.

[7] Pedrycz, W. (1997), *Fuzzy Sets Engineering*, CRC Press, Boca Raton, Fl.

INDEX